NEW APPLICATIONS OF ELECTRIC DRIVES

Edited by **Miroslav Chomat**

New Applications of Electric Drives
http://dx.doi.org/10.5772/60584
Edited by Miroslav Chomat

Contributors
Vu Trieu Minh, Mart Tamre, Alina Sivitski, Igor Penkov, Clément Pornet, Xi Zhang, Yunhua Li, Jun Xu, Binggang Cao, Tatiana Minav, Jani Heikkinen, Matti Pietola, Wenping Cao

CBS, Edition **2017**

Published by InTech
Janeza Trdine 9, 51000 Rijeka, Croatia

Publishing Process Manager Sandra Bakic
Technical Editor InTech DTP team
Cover InTech Design team

Additional hard copies can be obtained from orders@intechopen.com

New Applications of Electric Drives, Edited by Miroslav Chomat
p. cm.
ISBN 978-953-51-2233-3

INTECH

open science | open minds

Contents

Preface VII

Chapter 1 **Electric Drives in Alternative Fuel Vehicles — Some New Definitions and Methodologies 1**
Xi Zhang

Chapter 2 **Modeling and Control Strategy for Hybrid Electrical Vehicle 27**
Vu Trieu Minh, Alina Sivitski, Mart Tamre and Igor Penkov

Chapter 3 **Fault Diagnosis of Switched Reluctance Motors in Electrified Vehicle Applications 59**
Yihua Hu, Chun Gan, Wenping Cao and Stephen Finney

Chapter 4 **Battery Management System for Electric Drive Vehicles – Modeling, State Estimation and Balancing 87**
Jun Xu and Binggang Cao

Chapter 5 **Electric Drives for Propulsion System of Transport Aircraft 115**
Clément Pornet

Chapter 6 **Electric-driven Zonal Hydraulics in Non-Road Mobile Machinery 143**
Tatiana A. Minav, Jani E. Heikkinen and Matti Pietola

Chapter 7 **Variable Frequency Drive Applications in HVAC Systems 167**
Yunhua Li

Preface

The field of modern electric drives is very multidisciplinary and progress achieved in power electronics, electric machines, computer engineering, control sciences, informatics, or advanced materials have a significant impact on the overall improvements of the drives. As the electricity needed to power electric drives is, to an increasing extent, generated from renewable energy resources, the substitution of electric drives for internal combustion engines running mainly on fossil fuels, contributes to the reduction of greenhouse gas emissions. Excellent controllability of advanced variable-speed drives also enables achieving better properties of devices in which they are implemented.

Most of this book focuses on applications of electric drives in hybrid and electric vehicles, as one would expect. The problems of modeling, control, fault diagnostics, and battery management are particularly addressed. Applications in aircraft, non-road mobile machinery or in heating, ventilation and air-condition systems are also discussed. Apart from the description of the current state of the art in the respective fields, some new findings and methods are presented. The reader will find extensive bibliography mapping recent developments in the addressed areas, too.

I would like to thank all the participating authors for their valuable contributions and the InTech staff, particularly Ms. Sandra Bakic, the publishing process manager, for their professional approach to publishing this book.

Miroslav Chomat
Institute of Thermomechanics CAS, v.v.i.
Prague, Czech Republic

Chapter 1

Electric Drives in Alternative Fuel Vehicles — Some New Definitions and Methodologies

Xi Zhang

Additional information is available at the end of the chapter

http://dx.doi.org/10.5772/61645

Abstract

This chapter focuses on some new definitions and methodologies of electric drives that are facing new challenges raised by alternative fuel vehicles. It starts with the objectives, fundamentals, and current research issues of alternative fuel vehicles based electric drives, before moving on to new definitions of unified modeling of the entire electric drive system and design of the proposed DC active power filter aimed at energy storage system chaotic current elimination. Next, novel motor control strategies taking into account alternative fuel vehicle operations are presented for improvement of sensorless drive and flux weakening control performance. Finally, conclusions of this chapter are drawn.

Keywords: Alternative fuel vehicles, unified modeling, DC active power filter, sliding mode observer, flux weakening control

1. Introduction

1.1. Electric drive application background in Alternative Fuel Vehicles (AFVs)

The current worldwide energy shortage and unbalanced distribution is becoming increasingly serious, and resource competition is getting fierce. The potential petroleum resources gradually decline, while global oil consumption continues to increase. Traditional Automobiles have to face an unprecedented impact and challenges. On the other hand, with the traditional automobile production sharply increasing, environmental impact becomes more apparent. The emissions of CO_2 and other greenhouse gases and global warming [1] come into our view. Alternative fuel vehicles (AFVs) are gradually becoming an inevitable choice for all countries to innovate the automobile industry. For instance, in the United States, the sales volume of battery electric vehicles (BEVs) is expected to account for 12% of that of all registered vehicles in 2025. As an energy-efficient, low-carbon and environment friendly traffic tool, AFVs are expected to become increasingly popular in the future.

Most national research institutions and automobile manufacturers have increased R&D investment in the AFV technology to replace traditional vehicles, forming the trend of different technologies developed in parallel. Some of the technologies have achieved significant success in commercial fields [2], [3]. In the United States, Germany, and Japan, GM, Ford, Volkswagen, BMW, Toyota, Honda, and other major automobile manufacturers have adopted different new technology development strategies, based on the situation of the nation and the company, and successfully developed a number of concept AFVs and utility vehicles. AFVs are usually classified into hybrid electric vehicles (HEVs), electric vehicles (EVs), fuel cell electric vehicles (FCEVs).

Some countries, e.g., Japan, have been committed to the hybrid vehicles and have developed a lot of well-known models, e.g., the Toyota "Prius." Some other countries, e.g., Germany, have unique technology in the pure electric vehicles, fuel cell electric vehicles, and so on. BMW i3 with carbon fiber body is the lightest electric car on the market.

The world's major automobile production countries have policies to promote the development of AFVs to cope with difficulties and opportunities. By 2015, the United States raised the goal to one million plug-in hybrid electric vehicles (PHEVs). Tax incentives have been implemented. The EU will issue €7 billion loan to support energy-efficient vehicles. The German government raised million BEV and PHEV goals by 2020. Japan in 2009 proposed a "carbon revolution" program, in which development of electric vehicles is one of the core. In China, basic research and the establishment of public service platform are listed as the focus of development. "Three vertical and three horizontal" development layout is presented to promote the AFV industries.

1.2. Current research situation and new challenges for electric drives in AFVs

AFV electric drive systems include the energy storage systems (battery and/or ultracapacitor), electronic controller, power converters, motors, and so on. The electronic controller is divided into three functional units: sensors, circuits, and processors. In driving and regenerative processes, the energy flow between the energy source and the motor is adjusted by the power converter.

The electric drive system itself as a whole is closely linked with the energy subsystem, which is the core technology of electric vehicles. The main function of electric drive system is to transfer energy from storage system to wheels, which is used for overcoming rolling resistance, air resistance, acceleration resistance, and climbing resistance. When braking, the mechanical energy is converted to electrical energy back to the energy storage system. AFVs can choose single or dual motor drives. AFVs can also employ in-wheel motors. Now, low-power electric vehicles on the market almost all employ single-motor power systems with simple structure.

However, high-power electric vehicles such as electric buses generally use dual motor powered system. AC drives can be divided into induction motor drive system and synchronous motor drive system. Synchronous motor drive systems can be classified as permanent magnet synchronous motor drive system (PMSM Drives) and switched reluctance motor (SRM) drive system. Currently, the commercial application of power systems for electric vehicles is divided into three categories: the DC motor drive system, induction motor drive system, and permanent magnet synchronous motor drive system. Representative products include "Peugeot 106 EV" (DC motor), "General EVI" (induction motor), and "Honda EV Plus" (permanent magnet synchronous motor), respectively. With the development of power electronics, micro-electronics, and control technology, the digital AC drive systems (IM Drives and PMSM Drives) are widely applied in commercial AFVs.

A 3-phase 4-pole induction motor and an electronics module with 72 IGBTs are adopted by the famous electric car Tesla Model S. The efficiency of the battery pack, power electronics modules, and motor system can reach 85–95%. It weighs only 70 pounds, can generate large torque at very low speed, and the max motor speed could be 13,000 rpm, which cannot be done by the internal combustion engine.

At present, there are still disadvantages in the electric drive system. First, the long charging times, which usually exceed 30 minutes [4] for a full charge, as well as the range limitation of EVs due to the available battery technologies, are still challenging problems.

Short driving distance, greatly limits the application and development of electric vehicles. In electric vehicles, the power supply did no progress a lot. Effective use of the vehicle power supply energy is an urgent problem. Torque of the drive motor, speed control accuracy, and efficiency optimization requires improvement over the entire range.

The motor runs under bad conditions, e.g., the vehicle engine vibration and high-temperature environment. Therefore, the motor must be adapted to environmental conditions: reliable, stable, and safe operation. Due to space limitations, high-density, compact, and lightweight motors are in need.

1.3. Objectives and contents' overview

The objectives of this chapter are to analyse some new electric drive design requirements raised by dynamic driving cycles of AFVs; to introduce some novel ideas about integration modeling from the energy storage system (ESS) to motor and power electronic device design for ESS lifetime extension and to delineate a few latest control strategies of AFV traction motors related to sensorless drives and flux weakening control. The rest of this chapter is organized as follows. Section 2 describes the basic concepts of structure and main components of electric drives in AFVs. Integration of modeling and novel design of the DC active power filter for ESS lifetime extension are given in section 3. In section 4, advanced motor control strategies with respect to sensorless drives and flux-weakening regions are explained. Finally conclusions of this chapter are drawn in section 5.

2. Configurations and components of electric drives in AFVs

2.1. Electric drive configurations in various powertrains

2.1.1. Electric drives for pure electric powertrain with battery, ultracapacitor, or Hybrid Energy Storage System (HESS)

During the AFV running, the load profile consists of high current peaks and steep valleys due to the repetitive starting, acceleration, and braking in the urban driving cycle. Thus, the battery alone faces several challenges: 1) huge addition to the battery numbers to meet the peak power demand; 2) severe adverse effect on battery life because of frequent charge or discharge operations; 3) extra difficulty on battery thermal management in high-power-load situation [5].

The ultracapacitor has much higher power density, longer cycle life, lower temperature sensitivity, higher efficiencies, but lower energy density, which makes it an attractive pairing for batteries [6], [7].

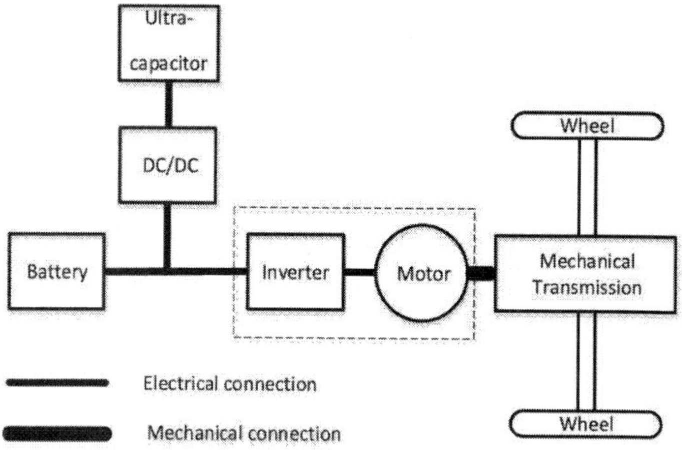

Figure 1. The configuration of hybrid energy storage system(HESS)-based electric vehicle

The topology of hybrid energy storage system is shown in Figure 1. Ultracapacitor and DC/DC converter are in series connection, with battery parallelly connected to the DC bus [8].

Hori [9], [10] adopted a heuristic control approach based on vehicle speed by taking into account ultracapacitors' energy utilization efficiency. Wang and Deng, et al. [8] adopted an optimization method for power distribution with the objective of minimum battery energy loss as the optimization goal. Ortuzar and Moreno [11] adopted heuristic algorithm based on two main rules and optimized algorithm obtained from a neural network as the energy management algorithms, respectively.

2.1.2. *Electric drives for extended-range/hybrid electric powertrain with participation of engines*

In a series HEV (SHEV), the internal combustion engine (ICE) is the main power source that converts the original energy in gasoline to mechanical power. The electric motor is actuated by the electricity generated by the generator or electricity stored in the battery, or both. The vehicle speed does not affect the engine speed because the engine and the wheels are decoupled. Thus, the engine can operate at its optimum speed to achieve the best fuel economy and be placed more flexibly.

A parallel hybrid electric vehicle (PHEV) adds a mechanical link (such as a clutch, belts, pulleys, and gears) between the engine and the final drive, so the engine can drive the wheels directly. The mechanical coupling adjusts the torque from the engine and the motor depending on the operation mode. The parallel hybrid electric vehicle has two propulsion devices, which can be used in motor-alone mode, combined power mode, engine-alone mode, power split mode, stationary charging mode, regenerative braking mode.

The series–parallel HEV can be operated as a series or parallel HEV. In comparison to a series HEV, the engine is coupled to the final drive through a mechanism, so the engine can deliver power to the final drive directly. In comparison to a parallel HEV, the extra mechanical output of the engine can be converted to electricity directly using the generator. The more complicated configuration brings out more flexible control strategy, better optimization of fuel economy and driveability, as well as higher cost.

2.1.3. *Fuel cell vehicle based electric drives*

The fuel cell powered hybrid drive train [12], [13] mainly contains a fuel cell system, a proton power system (PPS), an electric motor and accordingly motor control unit (MCU), a vehicle control unit (VCU), and a power converter between the fuel cell stack and the proton power system. After the accelerator, the braking pedal or other operating signals provide the power or torque command, the motor output power or torque is controlled by the VCU, as well as the energy flows between the fuel cell stack, the proton power system, and the drive train.

2.2. Motors, power converters, and electronic controllers of electric drives in AFVs

2.2.1. *Types and brief descriptions of traction motors*

Among different types of electric drives, there are four major types of motors that are viable for EVs: DC motors, induction motors, switched reluctance motor, and permanent magnet brushless motor. They possess fundamentally different machine topologies. Basically, they are classified into two main groups, the brushed and the brushless, and each group can be further classified into different subgroups.

2.2.1.1. *DC motor*

DC Motors are widely applied in EVs, because of the orthogonal disposition of field and armature. Simple control is the advantage of DC Motors. The permanent magnet DC Motors allow a significant reduction due to the efficient use of radial space. Due to the low permeability of the permanent magnet DC Motors, armature reaction decreases under normal circumstances, the commutation has been greatly improved. However, their commutator and brushes make operation not very reliable and maintenance is difficult.

2.2.1.2. *Induction motor*

Induction motors as a brushless drive are widely used for electric vehicles, because of their low cost, high reliability, and easy maintenance. However, the induction motor performance is quite limited, such as the variable voltage variable frequency (VVVF) control. The main reason is due to the dynamic model of the nonlinearities. Recently, the induction motor has seen a greater progress in the design method, and the power density is increased. On the other hand, efficiency-optimizing-control (EOC) approaches have been developed for the EV induction drives, which can reduce the consumed energy and increase the regenerative energy. Moreover, the program of changing the pole pair number has been developed to significantly extend the constant power region.

2.2.1.3. *Switched Reluctance Motor (SRM)*

SRMs with simple structure have been recognized as potential traction motors for EVs [14], low manufacturing cost, and outstanding torque speed characteristics. However, design and control are difficult and, in addition, acoustic-noise is still a problem. Now, fuzzy sliding mode control has been developed for the control of nonlinear systems and minimization of chattering. In addition, vibration cancellation technique for the SRM has been proposed to reduce the vibration mode and acoustic noise.

2.2.1.4. *Permanent Magnet Brushless Motor (PMBLM)*

Permanent magnet brushless motors (PMBLMs) are becoming increasingly attractive and can directly compete with the induction drives for EVs. The permanent magnet synchronous motor (PMSM) and brushless DC motor (BLDCM) are two main representatives. The advantages of PMBLMs are high efficiency, high power density, and high reliability. The key problem is the high cost due to PM materials. In recent years, the new class of PMBLMs has applied hybrid field excitations.

2.2.2. *Topologies of power converters and switching devices for AFVs' deployment*

An integral part of the electric vehicle comprises DC-AC inverters and DC-DC converters, used as an application for EV and HEV motor drives.

2.2.2.1. *DC-DC converter*

Owing to the advent of intelligent vehicle systems, the demand of DC power for automotive electronic equipment is continuously increasing. A DC-DC converter possesses the function of converting the input DC voltage to another output DC voltage with different levels. A high conversion efficiency can be usually guaranteed [15]. Normally, DC-DC converters are switching regulators with operations at high frequency. MOSFETs and IGBTs are used as switches. The size of the MOSFETs or IGBTs is drastically reduced at high-frequency operation. The switches are turned ON and OFF by pulse-width modulation (PWM) technique.

The applications of DC-DC converters have been implemented in AFVs. Bidirectional DC-DC converters are applied for battery charging, regenerative braking, and backup power. Unidirectional DC-DC converters are also used in DC motor drives and electric traction.

A bidirectional DC-DC converter as seen in Figure 2 can be divided into three main blocks. The primary side (low-voltage side) usually consists of a buck or boost-type converter and the secondary side is usually half or full-bridge arrangement.

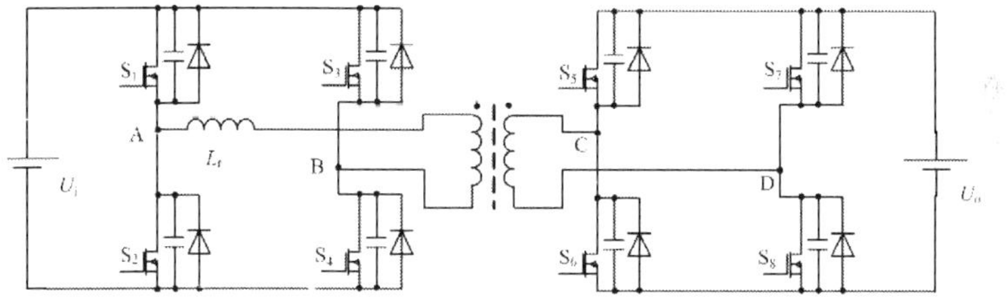

Figure 2. Full-bridge bidirectional DC-DC converter

2.2.2.2. *Inverter*

The DC–AC converter (inverter) has been the main choice in power electronics, because of its circuit simplicity and rugged control scheme. In order to expand the motor speed and torque region, usually high modulated frequency and high output current are permitted for the motor controlled by the inverter for AFV applications. Inverters improve the voltage utilization, reduce current harmonic contents, and create modulated sinusoidal voltage by sinusoidal PWM (SPWM) or space vector PWM (SVPWM) control methods. Three-phase two-level inverter, known as six switches inverter, shown in Figure 3, is widely used. This topology has the merits of low cost, easy control, and high reliability.

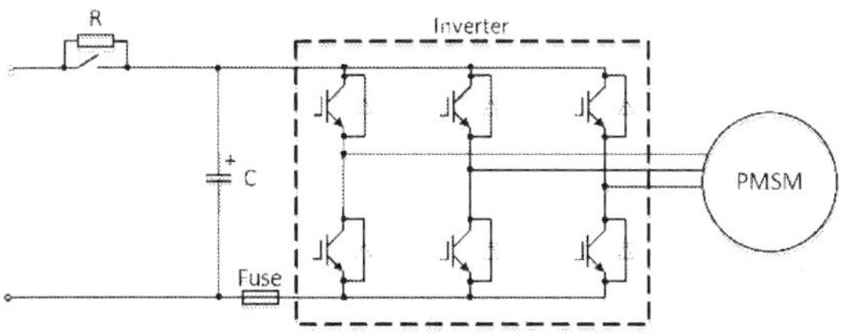

Figure 3. A common inverter topology

2.2.2.3. Soft-switching technology

Converters can be classified into two main categories: hard-switched and soft-switched for DC-DC converters and inverters. Soft-switching is a technology in which zero-voltage switching (ZVS) and/or zero-current switching (ZCS) of the switches are achieved mostly using the parasitic and/or additional components. Soft-switching reduces switching stresses (high voltage and current peaks) on the power devices during the turn-on and turn-off transients, which results in oversizing the safe operating area (SOA). Switching losses decrease for overlap size reduction of voltage and current. Severe *di/dt* and *dv/dt*, which produce electromagnetic interference (EMI), are also limited effectively [16].

2.2.2.4. Switching device

For the driving system, the high-power converter module is used. Converters control the high electric power. Insulated gate bipolar transistors (IGBTs) are used in the high-power modules as the switching devices. Other power electronics modules control the middle and low power, in which metal-oxide-semiconductor field effect transistors (MOSFETs) are used as the switching devices. The majority of devices used in power electronics are enhancement transistors, respectively, n-channel-enhancement power MOSFETs and IGBTs [17].

The main problem of the high-power modules is high electric power loss, which requires the liquid cooling system. For the future, power switching devices, very low on-resistance, and high-temperature operation are expected to simplify the cooling system of the converter. Silicon Carbide (SiC) or Gallium Nitride (GaN) devices are possible candidates. For the middle- and low-power modules, lower breakdown voltage is required. For these applications, high-frequency operation is expected, which leads to small capacitors and reactors.

2.2.3. Functions and design key points of Motor Control Units (MCUs)

1. The motor control unit (MCU) has main components:**Electronic controller**

The electronic controller consists of hardware circuit and control software. The hardware circuit mainly contains microprocessor and its minimum system, the detection circuit of motor current, voltage, speed, temperature, all kinds of hardware protection circuit, and the communication circuit (CAN Bus) with the external control unit like the vehicle controller and battery management system to exchange data. The control software realizes corresponding control algorithm depending on different types of motors. For instance, the strategies of flux weakening control, i_d=0 control and MTPA control are widely used in PMSM control software.

2. **Driver**

It converts the motor control signal which comes from microcontroller to PWM signal, which drives the power converter and implements the isolation of power signal and control signal.

3. **Power converter**

The power converter controls the motor current. The power devices that are widely used in electric vehicles contain the gate-turn-off thyristor (GTO), MOSFET, IGBT, and smart power module, etc.

http://dx.doi.org/10.5772/61645

During the development process, several key technologies have to be analyzed, such as the design of power circuits, heat sink, and the control strategies development as well as kinds of hardware and software protection functions.

3. New definitions of modeling and design of the DC APF for electric drives in AFVs

3.1. New definitions of mathematical modeling integration from ESS to motor

3.1.1. Purposes and meanings of multidomain coupling unified mathematical expressions for electric drives

In the past decade, plenty of literature has focused on the operation performance, parameter optimization, and control strategies of components in AFV electric drives. These researchers mostly use DC power supplies or electronic loads to simulate input/output characteristics of components, and meanwhile realistic influence between components (e.g., battery and traction motor) is seldom considered. To the author's best knowledge, there are no publications that regard the electric drive system as a whole for the systematic research on multidomain (dynamics, electromagnetics, electrochemistry, etc.) coupling mechanism.

The electric motors are required to operate in complicated conditions of frequent start/stop and acceleration/deceleration. High torque is needed at low speeds or when running uphill, while low torque is needed at high speeds. In the meantime, speed adjusting range should be very wide. This is significantly different from the industry applications where usually rated conditions are assigned to the electric motors. Therefore, the transient performance requirements for the AFV electric motors are very high and, further, are transferred to high transient requirements on the ESS through three-phase inverter. Additionally, the switching frequency of switching devices (e.g., IGBT) is high up to several or tens of kHz, the switching commutation time is extremely short, and the disturbance currents including diode recovery current exist, so the uncertain high-frequency transient chaotic currents should be produced across the ESS. The phenomenon will do harm to the stability and even lifetime of the ESS. Detailed descriptions can be found in section 3.2.1.

On the other hand, the ESS (battery for specific) AC impedance (including ohmic resistance, polarization resistance, capacitive reactance, Warburg impedance, and so on) characteristics will have impact on the transient performance of traction motors. The AC impedance characteristics of the battery determine the battery output power capability (supposing the electrochemical potential is a constant) and further influence the transient output power capability of the traction motor via the inverter. Consequently, the motor transient performance could be somehow restricted.

Thus, this section is aimed at importing new definitions of electric drive unified mathematical expressions for systematic exploration of transient multidomain coupling mechanism from dynamics-electromagnetics-electrochemistry based on the ESS, traction motor, inverter, and

DC-DC converter in the AFV electric drive system. Multiple subjects could be melted including energy storage, electric machine, and power electronics. The entire electric drive can be regarded as a whole without separating relationships between components, and a unified and general mathematic expression is capable of identifying specific numeric relations between transient requirements and parameters of components. In addition, the systematic-, circuit-, and component-class parameter optimization aimed at efficiency improvement in electric drives of alternative fuel vehicles could be achieved.

3.1.2. *Modeling framework delineation*

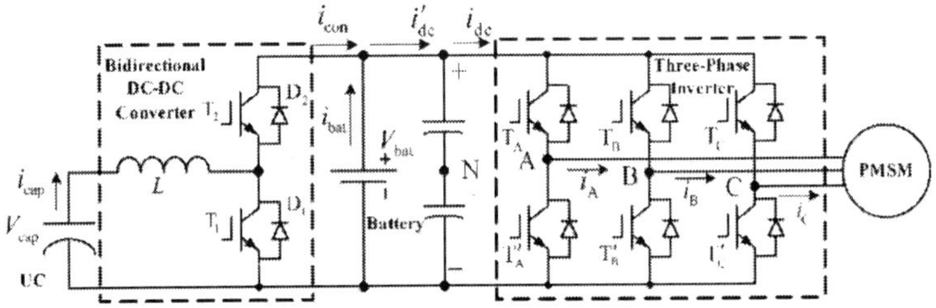

Figure 4. Circuit topology of the electric drive system

Considering comprehensiveness, an ESS including a battery and ultracapacitor is studied. The electric drive with only the battery is analyzed similarly and more easily. The circuit topology of the studied electric drive system in AFV is shown in Figure 4. The EV traction motor is a permanent magnet synchronous motor (PMSM) and the three-phase inverter is comprised of six IGBTs. The UC and battery are connected to the low-voltage (LV) and high-voltage (HV) sides respectively of the bidirectional DC-DC converter, which is constructed in a half-bridge topology.

Individual transient mathematical models (including the PMSM, inverter, DC-DC converter, ultracapacitor, and lithium-ion battery) of all components are built first. For instance, the switching state function is introduced to set up a nonlinear model of the inverter and then it is transferred to a linear model with help of the large-signal disturbance method. The lithium-ion battery is modeled using the scheme combining with the equivalent circuit and electro-chemical factor-assisted electrical parameter adaptive-adjusting.

The speed ω and torque T of the traction motor are chosen as unified inputs for the integrated systematic model. The individual model is transformed into expressions by $\omega(t)$ and $T(t)$. Thus, the multidomain models are related to each other using the unified inputs.

The important variables in all individual models can be finally expressed by $\omega(s)$ and $T(s)$, where s is a Laplace operator. The time-domain mathematical descriptions for these variables could also be acquired using combination of inverse Laplace transformation and some additional mathematical methods (e.g., residues method). Therefore, the multidomain coupling transient model for the AFV electric drives can be derived and complete frequency- and time-domain expressions of all variables as well.

3.2. Design of DC active power filter for ESS chaotic current elimination

3.2.1. Exploration of Interior relationship between traction motor output performance and ESS current ripples

The output A-phase voltage of the inverter can be expressed as its Fourier series expansions, as follows:

$$V_{\mathrm{AN}} = \sum_{k=0}^{+\infty} V_k \sin(k\omega t) \tag{1}$$

where V_k denotes the component of order k, ω represents the fundamental angular frequency. We can define $\alpha = 2\pi/3$ as the angular difference between two of three phases.

Regardless of the unbalanced currents problem, the nonsinusoidal phase currents can be depicted as follows:

$$i_{\mathrm{A}} = \sum_{m=0}^{+\infty} I_m \sin(m\omega t + \varphi_m), i_{\mathrm{B}} = \sum_{m=0}^{+\infty} I_m \sin(m(\omega t - \alpha) + \varphi_m), i_{\mathrm{C}} = \sum_{m=0}^{+\infty} I_m \sin(m(\omega t + \alpha) + \varphi_m) \tag{2}$$

The DC bus current can be considered as sum of currents through three bridges in the inverter. Meanwhile, one bridge's contribution is composed of the currents from upper and lower branches. Consequently, the DC bus current is obtained as:

$$i_{\mathrm{dc}} = i_{\mathrm{dc,A,1}} + i_{\mathrm{dc,A,2}} + i_{\mathrm{dc,B,1}} + i_{\mathrm{dc,B,2}} + i_{\mathrm{dc,C,1}} + i_{\mathrm{dc,C,2}} \tag{3}$$

where the subscript 1 represents the upper branch, and 2 represents the lower branch.

As an instance, the upper and lower currents for the Phase-A bridge are described as:

$$i_{\mathrm{dc,A,1}} = K_{\mathrm{A,1}} i_{\mathrm{A}}, i_{\mathrm{dc,A,2}} = K_{\mathrm{A,2}} i_{\mathrm{A}} \tag{4}$$

where the coefficients are $K_{\mathrm{A,1}} = \frac{V_{\mathrm{AN}}}{V_{\mathrm{bat}}} + \frac{1}{2}$, and $K_{\mathrm{A,2}} = \frac{V_{\mathrm{AN}}}{V_{\mathrm{bat}}} - \frac{1}{2}$. The Phase-B and Phase-C currents can be computed similarly. With expressions given by equations (1)-(4), the upper and lower currents of three bridges are available to be calculated.

Substituting equation (4) and similar expressions for the other two phases into (3), the DC bus current can be obtained in the form of Fourier series as:

$$
\begin{aligned}
i_{dc} = \sum_{m=1}^{+\infty} \frac{2I_m}{V_{bat}} \Bigg\{ & \sum_{k=1}^{+\infty} \frac{V_k}{2} \cos((k-m)\omega t - \varphi_m) \times [2\cos(k-m)+1] \\
& + \sum_{k=1}^{+\infty} \frac{V_k}{2} \cos((k+m)\omega t + \varphi_m) \times [2\cos(k+m)+1] \Bigg\}
\end{aligned} \tag{5}
$$

Finally, after combining similar terms and sequent calculation, the DC bus current expression is simplified to:

$$
i_{dc} = \sum_{m=1}^{+\infty} \frac{3I_m}{V_{bat}} \sum_{k=1}^{+\infty} (V_{3k+m} - V_{3k-m}) \cos(3k\omega t - \varphi_m) \tag{6}
$$

3.2.2. Overall descriptions for the novel DC active power filter

The operation principle of the proposed DC active power filter (APF) is delineated in Figure 5. The motor controller calculates the synchronous electrical angle and synchronous frequency of phase currents after acquiring rotor position and current/voltage signals. Thus, the fundamental frequency and the fundamental phase for the Fourier Transform of the DC bus current are determined with respect to the theoretical analysis in section 3.2.1. It has to be noted that every Fourier Transform decomposition period could give its information to the current allocation of the next period. So it is available to be predicted that the resulted low-frequency components (closely related to the PMSM speed) cannot change in such a short time. The high-frequency components are derived from the difference of the real-time samplings of the DC bus current and low-frequency components calculated in the previous processing period.

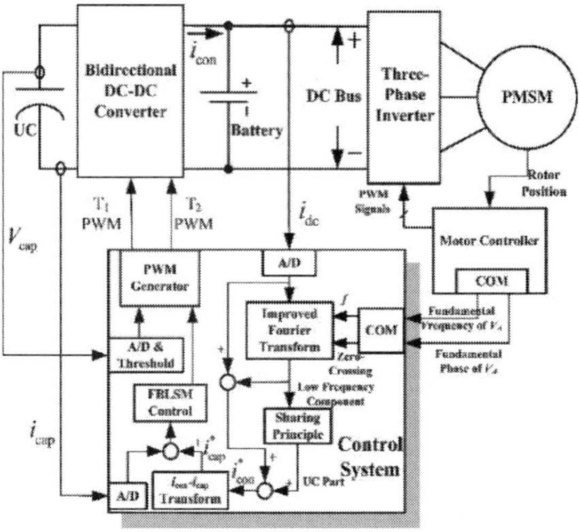

Figure 5. Operation principle of the proposed APF strategy with help of motor information.

In the peak power demand process (e.g., acceleration, uphill), the UC is commanded as a large power provider. Thus, the low-frequency components from the DC bus current should be shared reasonably by the UC and battery considering their respective power density. In this paper, the current directly from the UC is considered as a control variable; thereafter, the output current reference i_{con}^* on the DC-DC converter high-voltage side has to be transformed to the counterpart i_{cap}^* on the low-voltage side, complying with the power conservation law and estimated efficiency. The fixed-boundary-layer sliding mode (FBLSM) control is applied to achieve pulse-width-modulation (PWM) duty cycle for one of two IGBTs (choose T1 or T2 regarding power flow direction) due to consideration of eliminating of control chattering usually caused by frequent control switching in normal sliding mode control. In the meantime, the UC discharging stops when UC SOC reaches the desired lower limit, and then the battery or regenerative braking UC immediately helps UC increase its SOC to a preset value for the further use.

3.2.3. Improved fast fourier transform for frequency components' decomposition of ESS currents

The Discrete Fourier Transform (DFT) and Inverse Discrete Fourier Transform (IDFT) are expressed as the following two equations, respectively:

$$X(k)=\sum_{n=0}^{N-1}x(n)e^{-j\frac{2\pi kn}{N}}=\sum_{n=0}^{N-1}x(n)W_N^{kn},\quad 0\le k\le N-1 \tag{7}$$

$$x(n)=\frac{1}{N}\sum_{k=0}^{N-1}X(k)e^{j\frac{2\pi kn}{N}}=\sum_{k=0}^{N-1}x(k)W_N^{-kn},\quad 0\le n\le N-1 \tag{8}$$

In order to speed up the computation, the normal FFT is used. The FFT calculation is finally transformed to only two computations by using the following expressions:

$$X(k)=G(k)+W_N^kH(k),\quad 0\le k\le\frac{N}{2}-1 \tag{9}$$

$$X\left(k+\frac{N}{2}\right)=G(k)-W_N^kH(k),\quad 0\le k\le\frac{N}{2}-1 \tag{10}$$

where $G(k)=\sum_{r=0}^{N/2-1}x(2r)W_{N/2}^{rk}$ and $H(k)=\sum_{r=0}^{N/2-1}x(2r+1)W_{N/2}^{rk}$

To minimize the side lobes, the Blackman-Harris Window as a generalization of the Hamming family, is given by:

$$w(n)=\sum_{k=0}^{3}(-1)^k a_k\cos\left(\frac{2\pi kn}{N}\right) \tag{11}$$

Therefore, the improved FFT expression with the Blackman-Harris Window can be shown as follows based on equations (9) and (10):

$$X_w(k) = \sum_{i=0}^{3} (-1)^i \frac{a_i}{2}\left[X(k-i) + X(k+i)\right] \tag{12}$$

With use of the improved FFT, the high-frequency and low-frequency components can be decomposed from the load demand, for the power distribution to UC and battery, as shown in Figure 5.

4. Advanced motor control strategies for AFV-based electric drives

4.1. Overview of traction motor speed/torque control strategies

4.1.1. Speed/torque vector control

The concept of vector control was first proposed by Siemens' F. Blaschke in the early 1970s, which is also named as field-oriented control (FOC). In this method, the stator current of a three-phase AC motor is transformed into two orthogonal components [18]. One component can control the magnetic flux; the other can control the torque.

In a typical FOC block diagram for a permanent magnet synchronous motor (PMSM), the control system first transforms the command speed or torque into the corresponding current command i_d^*, i_q^*. Then, proportional-integral (PI) controllers are used as current regulator to generate the reference d–q axis reference voltage. The transistor switching is defined by the pulse-width modulation signal according to the stator references voltage.

For the FOC system, the most important thing is to realize the Clarke and Park transformation. It is these transformations that connect the three-phase coordinate system to the $d-q$ axis coordinate system. The Clarke transformation converts three-phase currents to a two-axis plot to create time-varying quadrature-current values i_α and i_β. And the Park transformation is between the stationary coordinates $\alpha-\beta$ and the $d-q$ coordinate reference frame rotated to align with the rotor flux. As the d axis is always in the direction of magnet flux linkage and q axis in the quadrature direction, the values of i_d and i_q remain constant. The flux and torque are separately controlled by stator direct-axis current i_d and quadrature-axis current i_q.

FOC is used to control the AC synchronous and induction motors by transforming their dynamic structure into that of DC machines. Unfortunately, the scheme requires many electronic components such as sensors, amplifiers, or multipliers which increase the cost and more calculations are needed to realize the control strategy. This situation is changed due to the availability of microelectronics, with the cost of control hardware being no longer a major consideration [19].

http://dx.doi.org/10.5772/61645

4.1.2. Direct Torque Control (DTC)

Direct torque control (DTC) is a method that controls the torque (and thus finally the speed) of three-phase AC electric motors. The key point of this control is to select stator voltage vectors according to the differences between the references of torque and stator flux linkage and their actual values estimated on basis of the measured voltage and current of the motor [20], [21].

In a PMSM drive with DTC, the *dq* axis currents i_d and i_q, are obtained from the measured three-phase currents, and the voltages v_d and v_q are calculated from the dc-link voltage, since the voltage vectors determined by the switching table are known. By integrating the stator voltages, we can calculate the stator flux linkage. Torque is calculated by a cross product of calculated stator flux linkage vector and measured motor current vector. Then the calculated flux magnitude and torque are compared with their reference values to get the flux and torque errors. If either the calculated flux or torque deviates from the reference more than the allowed tolerance, the transistors of the variable frequency drive are turned off and on to generate one certain voltage.

Without the current controller followed by a pulse-width modulation (PWM) comparator, and not using the parameters of the motor, except the stator resistance, the DTC owns the advantages of less parameter dependence and fast torque response when compared with the torque control via PWM current control. But torque and flux linkage ripples are always the main disadvantage of the DTC.

4.2. Sensorless drives for Permanent Magnet Synchronous Motors (PMSMs) in AFVs

Due to the characteristic of getting rid of sensors and reduction of cost and systematic complexity, the sensorless drives for AFV motors have attracted great interest of the auto industry. In a sensorless drive, the motor speed could be accurately estimated through some sophisticated observer and some other electrical parameters (currents, voltages, or flux) as well. Sliding mode control has been widely used for motor control systems because of its insensitive characteristics to disturbances of motor parameters. With the same property, sliding mode observers (SMOs) have been adopted by researchers and the industry in the last decade applied to the motor drives for precise estimation of speed and flux. It is more robust to the operating conditions and systematic uncertainties. However, it has to be noted that imperfect implementation of the switching leads to chattering, which is a major drawback of SMO. Thus the following section will introduce a new fixed-boundary-layer sliding mode observer for AFV motor drives, considering reduction of chattering and guaranteeing estimation precision in a wide speed range.

4.2.1. Mathematical model of studied motor

The induction motor is studied as an example for the explanation of the proposed sliding mode observer. Its mathematical model could be described in the rotating d-q frame [22] as follows.

Let i_d and i_q, V_d andV_q, ω_r, T_L , ψ_{rd}, θ, , p, and J stand for stator currents, stator voltages, rotor speed, load torque, direct-axis rotor flux, flux angle, pole pair number, and moment of inertia.

$\tau_r = L_r / R_r$ represents the rotor time constant, and $\sigma = 1 - L_m^2 / L_s L_r$ is the leakage magnetic coefficient. For analysis simplicity, set $m = (L_r R_s + L_m^2 R_r) / (\sigma L_s L_r^2)$, $\gamma = L_m / (\sigma L_s L_r)$, $\varsigma = 1 / (\sigma L_s)$, $\kappa = 1 / \tau_r$, $\rho = pL_m / L_r$, and the motor model can be written below:

$$\begin{cases} \dfrac{di_d}{dt} = -mi_d + \kappa\gamma\psi_{rd} + p\omega_r i_q + \kappa\dfrac{L_m}{\psi_{rd}} i_q^2 + \varsigma V_d \\ \dfrac{di_q}{dt} = -mi_q - \gamma p\omega_r\psi_{rd} - p\omega_r i_d - \kappa\dfrac{L_m}{\psi_{rd}} i_d i_q + \varsigma V_q \\ \dfrac{d\theta}{dt} = p\omega_r + \kappa\dfrac{L_m}{\psi_{rd}} i_q \\ \dfrac{d\psi_{rd}}{dt} = -\kappa\psi_{rd} + \kappa L_m i_d \\ \dfrac{d\omega_r}{dt} = \rho\psi_{rd} i_q - \dfrac{1}{J} T_L \end{cases} \tag{13}$$

4.2.2. *Design of fixed-boundary-layer sliding mode observer*

In this study, the three-phase stator currents are the only required measures, and these are transformed from the three-phase reference frame to a diphasic reference frame, and then to the frame of the rotating field (d-q) as follows:

$$\begin{cases} i_d = \sqrt{\dfrac{2}{3}}\left(\cos(\hat{\theta})i_a + \cos(\hat{\theta} - \dfrac{2}{3}\pi)i_b + \cos(\hat{\theta} + \dfrac{2}{3}\pi)i_c\right) \\ i_q = \sqrt{\dfrac{2}{3}}\left(-\sin(\hat{\theta})i_a - \sin(\hat{\theta} - \dfrac{2}{3}\pi)i_b - \sin(\hat{\theta} + \dfrac{2}{3}\pi)i_c\right) \end{cases} \tag{14}$$

where i_a, i_b, and i_c represent the three-phase stator currents, respectively, and $\hat{\theta}$ denotes the estimated flux angle. Relation between the estimated flux angle and rotor speed could be derived from system (13) as follows:

$$\frac{d\hat{\theta}}{dt} = p\hat{\omega}_r + \kappa\frac{L_m}{\hat{\psi}_{rd}} i_q \tag{15}$$

Assume that system (13) has the outputs $(y_1 \ \ y_2)^T = (i_d \ \ i_q)^T$. Unfortunately, it could be found that ψ_{rd} in system (13) is not observable. Reference [24] has proved that system (13) has a stable zero dynamics about ψ_{rd} with the output $y_1 = i_d$ and $\kappa > 0$. Thus, it is possible to estimate ψ_{rd} by designing an estimator, and then ω_r becomes observable. The sliding mode observer technology is employed to estimate ω_r. The required voltage d-q frame in the observer is obtained from outputs of voltage feedforward compensation part. The description of the fixed-boundary-layer sliding mode (FLBSM) observer is given below.

$$\begin{cases} \dfrac{d\hat{i}_d}{dt} = -mi_d + B_1\kappa\gamma\hat{\psi}_{rd} + B_1\lambda_1(i_d - \hat{i}_d) + B_1\kappa\dfrac{L_m}{\psi_{rd}}i_q^2 + \varsigma V_d \\ \dfrac{d\hat{\psi}_{rd}}{dt} = -\kappa\hat{\psi}_{rd} + \kappa L_m i_d \\ \dfrac{d\hat{\omega}_r}{dt} = B_1\rho\hat{\psi}_{rd}i_q + B_2\lambda_2\text{sat}((\tilde{\omega}_r - \hat{\omega}_r),\phi_1) \\ \hat{T}_L = -B_3J\lambda_2\text{sat}((\tilde{\omega}_r - \hat{\omega}_r),\phi_1) \end{cases} \tag{16}$$

where $\tilde{\omega}_r = \dfrac{B_2\lambda_1\text{sat}((i_d - \hat{i}_d),\ \phi_2)}{pi_q}$ denotes the auxiliary state of the rotor speed positioned where the speed measurement originally stays for observer design, $\hat{i}_d$ is an intermediate variable required by the observer to obtain estimation of other parameters so that i_d could be measured, and the function sat(x, ϕ) is defined below:

$$\text{sat}(x,\phi) = \begin{cases} x/\phi & \text{if } |x| < \phi \\ \text{sgn}(x) & \text{otherwise} \end{cases} \tag{17}$$

The coefficients B_1, B_2, and B_3 in observer (16) are depicted as follows:

$$B_1 = \begin{cases} 0, & \text{if } |\hat{\psi}_{rd} - \psi_{rd}| > \varepsilon \\ 1, & \text{otherwise} \end{cases} \tag{18}$$

$$B_2 = \begin{cases} 0, & \text{if } \hat{i}_d \neq i_d \\ 1, & \text{otherwise} \end{cases} \tag{19}$$

$$B_3 = \begin{cases} 0, & \text{if } \hat{\omega}_r \neq \tilde{\omega}_r \\ 1, & \text{otherwise} \end{cases} \tag{20}$$

Then, the theorem that the sliding mode observer can exponentially converge will be proved. Assume that the observation error of ψ_{rd}, $e_1 = \psi_{rd} - \hat{\psi}_{rd}$, exceeds ε (i.e., $B_1=0$), and $de_1/dt = -\kappa e_1$ where $\kappa>0$ can be obtained. A Lyapunov function $V_{e_1}=0.5e_1^2$ is designed. Then $dV_{e_1}/dt = -\kappa e_1^2 < -\kappa\varepsilon^2 < 0$ can be obtained. The exponential convergence of $\hat{\psi}_{rd}$ to ψ_{rd} is proved. Therefore, a certain instant t_1 can always be found such that when $t>t_1$, we have $e_1 \leq \varepsilon$ given a sufficiently small real $\varepsilon>0$. At this moment, $B_1=1$. Next $e_2 = i_d - \hat{i}_d$ is defined. When $t>t_1$, the following equation can be derived as:

$$\frac{de_2}{dt} = p\omega_r i_q + \left(\kappa\gamma - \frac{\kappa L_m i_q^2}{\psi_{rd}\hat{\psi}_{rd}}\right)e_1 - \lambda_1 \text{sat}(e_2, \phi_2) \tag{21}$$

To prove the convergence of $\hat{i}_d$ to i_d, a Lyapunov function $V_{e_2} = \frac{1}{2}e_2^2$ is used. We set

$$\lambda_1 = \max\left\{\left(\kappa\gamma - \frac{\kappa L_m i_q^2}{\psi_{rd}\hat{\psi}_{rd}}\right)e_1 + p\omega_r i_q\right\} + \xi_2 \tag{22}$$

and the differential of the Lyapunov function can be delineated as:

$$\frac{dV_{e_2}}{dt} = \left(\left(\kappa\gamma - \frac{\kappa L_m i_q^2}{\psi_{rd}\hat{\psi}_{rd}}\right)e_1 + p\omega_r i_q - \lambda_1 \text{sat}(e_2, \phi_2)\right)e_2 \le -\xi_2 |e_2| = -\sqrt{2V_{e_2}}\xi_2 \tag{23}$$

Consequently, $\hat{i}_d$ converges to i_d in a finite time t_2. When $t > t_2$, $B_1 = 1$ and $\frac{de_2}{dt} = e_2 = 0$ are obvious. So according to (21), the new relation establishes below:

$$p\omega_r i_q + \left(\kappa\gamma - \frac{\kappa L_m i_q^2}{\psi_{rd}\hat{\psi}_{rd}}\right)e_1 - \lambda_1 \text{sat}(e_2, \phi_2) = 0 \tag{24}$$

We set $E = \left(\kappa\gamma - \frac{\kappa L_m i_q^2}{\psi_{rd}\hat{\psi}_{rd}}\right)e_1 / p i_q$, so the auxiliary state of rotor speed is given by:

$$\tilde{\omega}_r = \omega_r + \left(\kappa\gamma - \frac{\kappa L_m i_q^2}{\psi_{rd}\hat{\psi}_{rd}}\right)e_1 / p i_q = \omega_r + E \tag{25}$$

E and $\frac{dE}{dt}$ exponentially converge to zero due to the convergence of the rotor flux. Thus, the fact that $\tilde{\omega}_r$ exponentially converges to ω_r is proved. Then we have:

$$\frac{d\tilde{\omega}_r}{dt} = \frac{d\omega_r}{dt} + \frac{dE}{dt} = \rho\psi_{rd} i_q - \frac{1}{J}T_L + \frac{dE}{dt} \tag{26}$$

A Lyapunov function $V_{e_3} = \frac{1}{2}e_3^2$ is used for proof of convergence of $\hat{\omega}_r$ to $\tilde{\omega}_r$ assuming $e_3 = \tilde{\omega}_r - \hat{\omega}_r$. Let

$$\lambda_2 = \max\left\{\rho i_q e_1 - \frac{1}{J}T_L + \frac{dE}{dt}\right\} + \xi_3 \tag{27}$$

given $\xi_3>0$. Then, the following equation can be obtained:

$$\frac{dV_{e_3}}{dt} = \left(\rho i_q e_1 - \frac{1}{J}T_L + \frac{dE}{dt}\right)e_3 \le -\xi_3|e_3| = -\sqrt{2V_{e_3}}\xi_3 \tag{28}$$

Therefore, the convergence of $\hat{\omega}_r$ to $\tilde{\omega}_r$ in a finite time t_3 is proved. When $t>t_3$, $B_3=1$and $\frac{de_3}{dt}=e_3=0$ are obvious. Thus,

$$\hat{T}_L = T_L - J\rho i_q e_1 - J\frac{dE}{dt} \tag{29}$$

As mentioned above, both $\tilde{\omega}_r$ and $\hat{\psi}_{rd}$ exponentially converge, so e_1, E, and dE/dt exponentially converge to zero. Therefore, the convergence of $\hat{T}_L$ to T_L is finally proved.

The reference, experimental, and estimated values of rotor speed of the electric motor are compared in Figure 6. To test the robustness of the observer, +100% resistance change is considered and the -50% load torque and +100% stator leakage inductance changes as well. We can see that the actual speed tracks the reference trajectory well using the proposed controller, and no distinct speed differences between estimation and measurement occurred. Because of small inaccuracy due to the current harmonics, differences between the estimated and measured speed at low speeds seem a little bit bigger than those at high speeds. However, the situation is also acceptable in real operation.

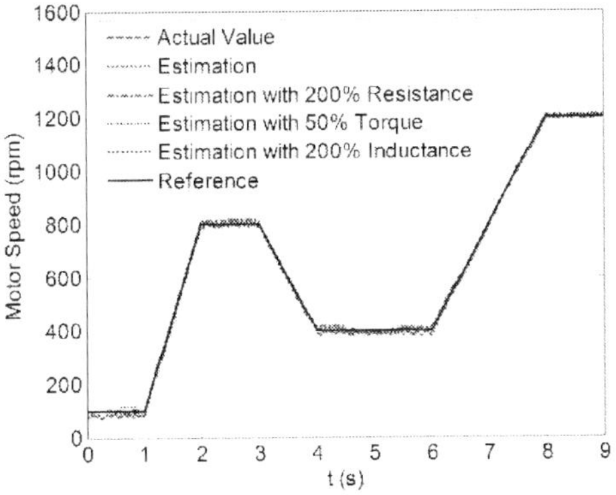

Figure 6. Comparison between references and estimated speeds using FBLSM observer

Rotor flux comparison is also studied, given the flux reference. Unlike the rotor speed results, estimated rotor flux is not precisely approaching the reference (see Figure 7), which is however close to actual results. The "actual value" shown in Figure 7 for comparison is obtained from normal flux calculation method through measured voltages and currents and could be considered as actual flux due to proof of incomputable applications. It is clear that the estimated flux with +200% resistance, 50% load torque, or +200% stator leakage inductance is not influenced by these disturbances dramatically.

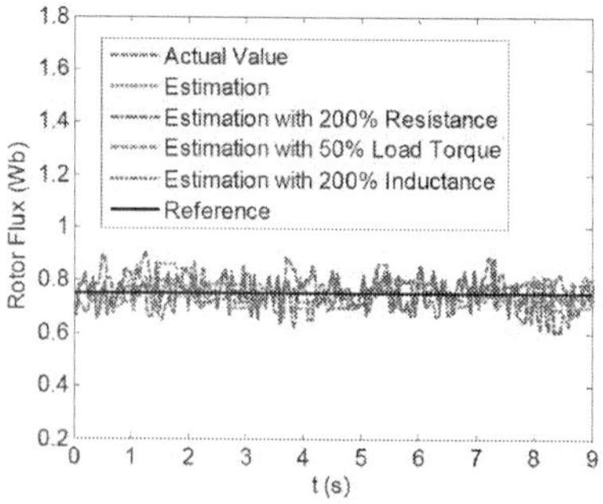

Figure 7. Comparison between references and estimated rotor flux using FBLSM observer

4.3. Advanced flux-weakening control at high speeds required by AFE driving

In AFVs, the traction motor has a very wide speed range, which needs flux-weakening control strategy to make the motor work at high speeds.

4.3.1. Classification of traction motor operation ranges

Figure 8 illustrates the characteristics of a permanent magnet synchronous motor used in AFVs. All the operation ranges can be divided into three parts: constant torque, constant power, and high-speed region. When the rated speed is not reached, the PMSM works in the constant-torque region, exerting a constant torque (rated torque). Once the speed reaches the rated motor speed, the torque begins to drop proportionally compared with speed, leading to a constant output power. The constant-power region ends when the following high-speed region starts. Then with the square of the speed, the motor torque drops proportionally [23].

4.3.2. Depictions of condition identification for flux-weakening control

Permanent magnet synchronous motors (PMSMs) fed by PWM inverters are widely used for industrial applications, especially servo drive applications, which always work in rated condition. But in traction and spindle drives, higher speed is needed [24]. When in constant-power region, the motor is always controlled by the flux-weakening control method.

In the constant torque region, the stator voltage increase as the internal EMF of the motor rises in proportion to the motor speed. However, when the speed rises to rated speed, the voltage

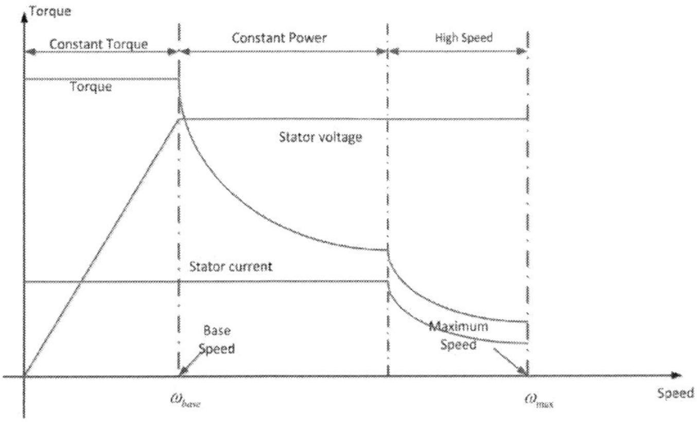

Figure 8. Characteristic curve of PMSM

limit of the associated frequency converter is reached. The internal voltage must now be adjusted to be compatible with the applied converter voltage. However, since the permanent magnets inherently provide the equivalent of a constant field excitation, the internal EMF of the machine continues to increase as speed increases. So the air gap flux should be weakened by the demagnetizing effect due to the d-axis armature reaction, which is called flux weakening [25].

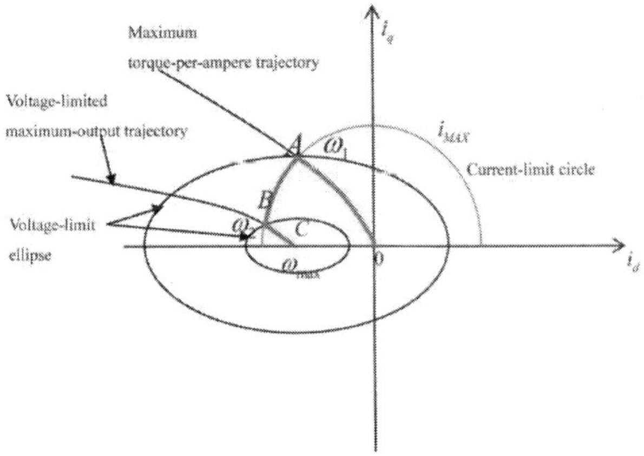

Figure 9. $i_d - i_q$ plane

On the $i_d - i_q$ plane (see Figure 9), there are a current-limit cycle and voltage-limit ellipse. These two limits are all decided by the inverter capacity. The maximum torque-per-ampere current vector trajectory is used in the constant torque region to generate maximum torque. The voltage-limited maximum output trajectory makes the motor increase to higher speed.

Region I ($\omega \leq \omega_1$): i_d and i_q are constant values given by the maximum torque-per-ampere (MTPA) trajectory. The maximum output torque is usually achieved in point A in Figure 8, where stator current $I_s = I_{\text{lim}}$, stator voltage $V_s < V_{\text{lim}}$

Region II ($\omega_1 < \omega \leq \omega_2$): when the speed reaches ω_1 and stator voltage increases to maximum voltage, the current vector starts to moves from A to B along the current limit circle as the rotor speed increases. In this region, stator current $I_s = I_{\text{lim}}$, stator voltage $V_s = V_{\text{lim}}$

Region III ($\omega_2 < \omega$): when at point B, $\omega = \omega_2$, the current vector starts to move along the voltage-limited maximum output trajectory, where stator current $I_s < I_{\text{lim}}$, stator voltage $V_s = V_{\text{lim}}$ [26].

4.3.3. Proposal of novel flux-weakening control based on nonlinear current compensation

Figure 10(a) illustrates the block diagram of the novel flux-weakening control based on nonlinear current compensation. Part 1 is the torque controller in the constant torque region. The current command i_s^* is the output of the P1I speed controller, and then decomposed into d- and q-axes components, i_{dx}^* and i_{qx}^* according the maximum torque-per-ampere trajectory until the current regulator begins to saturate.

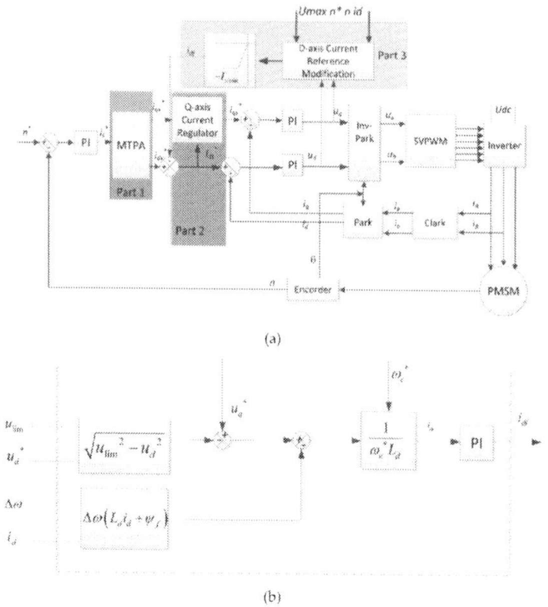

Figure 10. (a) The block diagram of the novel flux-weakening control based on nonlinear current compensation. (b) d-axis current modification

The proposed flux-weakening controller includes Parts 2 and 3. The proposed flux-weakening control algorithm uses the output reference voltage of the synchronous PI current regulator to identify the onset of the flux weakening. As the speed of the IPMSM increases, the output of the current regulator approaches to the boundary of the pulse-width modulator. With the reduced margin of the voltage, the performance of the current regulator gets worse and finally loses its controllability. As shown in Figure 10(b), the controller ensures the margin of the voltage by compensating the d-axis current toward the negative direction to prevent saturation of the current regulators. But at the low-speed region, the output voltage of the current regulator is usually less than V_{max} and thus Part 3 is not activated. The function of Part 2 is to adjust the q-axis current command i_{qx}^*, as the d-axis current command i_{qs}^* increased toward the negative direction, because of the current limit I_{smax}.

5. Conclusions

In this chapter, some fundamentals of the electric drives in alternative fuel vehicles are first introduced including application background and existing challenges. Then the multidomain transient unified modeling of AFV electric drives is newly defined at aim of efficiency improvement and ESS lifetime extension. The active power filter (APF) in the DC system is proposed for the elimination of ESS chaotic currents. The sensorless drive for the traction motor using a fixed boundary layer sliding mode observer is described in detail, as well as the flux-weakening controller using nonlinear current compensation, in order to enhance traction motor control precision and dynamic performance. The definitions and strategies discussed in this chapter may contribute to research and development of brand-new electric drives applied in EVs or HEVs.

Acknowledgements

The author would like to thank the National Natural Science Foundation of China for the support under Grant 51277121.

Author details

Xi Zhang

Address all correspondence to: braver1980@sjtu.edu.cn

Institute of Automotive Engineering, Shanghai Jiao Tong University, Shanghai, China

References

[1] Paiva SLD, Savi MA, Viola FM, Leiroz AJK. Global warming description using Daisyworld model with greenhouse gases. *Biosystems*. 2014;125(2014):1-15. DOI: 10.1016/J.BIOSYSTEMS.2014.09.008

[2] Zhang L, Brown T, Samuelsen S. Evaluation of charging infrastructure requirements and operating costs for plug-in electric vehicles. *J Power Sources*. 2013;240(2013): 515-524. DOI: 10.1016/J.JPOWSOUR.2013.04.048

[3] Rangarajua S, Vroeyb LD, Messagiea M, Mertensb J, Mierloa JV. Impacts of electricity mix, charging profile, and driving behavior on the emissions performance of battery electric vehicles: A Belgian case study. *Appl Energy*. 2015;148(2015):496-505. DOI: 10.1016/J.APENERGY.2015.01.121

[4] Neubauer J, Wood E. The impact of range anxiety and home, workplace, and public charging infrastructure on simulated battery electric vehicle lifetime utility. *J Power Sources*. 2014;257(2014):12-20. DOI: 10.1016/J.JPOWSOUR.2014.01.075

[5] Carreira D, Domingos Marques G, Sousa DM. Hybrid energy storage system joining batteries and supercapacitors. 2014 IEEE 5th International Symposium on Power Electronics for Distributed Generation Systems (PEDG); Jun 24-27; Galway. 2014. p. 1-6. DOI: 10.1109/PEDG.2014.6878712

[6] Doucette RT, McCulloch MD. A comparison of high-speed flywheels, batteries, and ultracapacitors on the bases of cost and fuel economy as the energy storage system in a fuel cell based hybrid electric vehicle. *J Power Sources*. 2011;196(3):1163-1170. DOI: 10.1016/J.JPOWSOUR.2010.08.100

[7] Li Q, Chena W, Li Y, Liu S, Hua J. Energy management strategy for fuel cell/battery/ultracapacitor hybrid vehicle based on fuzzy logic. *Int J Electric Power Energy Sys*. 2012;43(1):514-525. DOI: 10.1016/J.IJEPES.2012.06.026

[8] Wang T, Deng W, Wu J, Zhang Q. Power optimization for hybrid energy storage system of electric vehicle. Transportation Electrification Asia-Pacific (ITEC Asia-Pacific), 2014 IEEE Conference and Expo; Aug 31-Sep 3; Beijing, China. 2014. p. 1-6. DOI: 10.1109/ITEC-AP.2014.6941224

[9] Guidi G, Undeland TM, Hori Y. Effectiveness of supercapacitors as power-assist in pure EV using a sodium-nickel chloride battery as main energy storage. EVS24; May 13-16; Stavanger, Norway. 2009. p. 1-9.

[10] He X, Parten M, Maxwell T, Energy management strategies for a hybrid electric vehicle. IEEE Conference on Vehicle Power and Propulsion; Sep 7-9; Chicago, USA. 2005. p. 536-540. DOI: 10.1109/VPPC.2005.1554610

[11] Ortuzar M, Moreno J, Dixon J. Ultracapacitor-based auxiliary energy system for an electric vehicle: implementation and evaluation. *IEEE Trans Indus Electron*. 2007;54(4): 2147 - 2156. DOI: 10.1109/TIE.2007.894713

[12] Cipollone R, Battista DD, Marchionni M, Villante C. Model based Design and Optimization of a Fuel Cell Electric Vehicle. *Energy Procedia*. 2014;45(2014):71-80. DOI: 10.1016/J.EGYPRO.2014.01.009

[13] Ansarey M, Panahib MS, Ziarati H, Mahjoob M. Optimal energy management in a dual-storage fuel-cell hybrid vehicle using multi-dimensional dynamic programming. *J Power Sources*. 2014;250(2014):359-371. DOI: 10.1016/J.JPOWSOUR.2013.10.145

[14] Woothipatanapan S, Jangwanitlert A, Chancharoensook P. Efficiency improvement of converter for switched reluctance motor drives by mixed parallel operation of IGBT and MOSFET. TENCON 2010 - 2010 IEEE Region 10 Conference; Nov.21-24; Fukuoka, Japan. 2010. p. 1841-1846. DOI: 10.1109/TENCON.2010.5685964

[15] Christen D, Tschannen S, Biela J. Highly efficient and compact DC-DC converter for ultra-fast charging of electric vehicles. 2012 15th International Power Electronics and Motion Control Conference (EPE/PEMC); Sep 4-6; Novi Sad, Serbia. 2012. p. LS5d. 3-1-LS5d.3-8. DOI: 10.1109/EPEPEMC.2012.6397481

[16] Bellar MD, Wu TS, Tchamdjou A, Mahdavi J, Ehsani M. A review of soft-switched DC-AC converters. *IEEE Trans Indus Applic*. Jul-Aug 1998;34(4):847-860. DOI: 10.1109/28.703992

[17] Tulbure A, Turschner D, Abrudean M, Ceuca E, Ormenisan R. Experimental comparation of switching with IGBT and MOSFET. Automation Quality and Testing Robotics (AQTR), 2010 IEEE International Conference on. 28-30 May 2010;2:1-5. DOI: 10.1109/AQTR.2010.5520804

[18] Abu-Rub H, Iqbal A, Guzinski J. Vector control of three-phase AC machines: *System Development in the Practice*. 1st ed. Verlag, Germany: Springer; 2008. 340 p. DOI: 10.1007/978-3-540-79029-7

[19] Dominic DA, Chelliah TR. Analysis of field-oriented controlled induction motor drives under sensor faults and an overview of sensorless schemes. *ISA Trans*. 2014;53(5):1680-1694. DOI: 10.1016/J.ISATRA.2014.04.008

[20] Rahman MF, Zhong L, Lim KW. A Direct torque-controlled interior permanent magnet synchronous motor drive incorporating field weakening. *IEEE Trans Indus Applic*. Nov/Dec 1998;34(6):1246-1253. DOI: 10.1109/28.738985

[21] Zhong L, Rahman MF, Hu WY, Lim KW. Analysis of direct torque control in permanent magnet synchronous motor drives. *IEEE Trans Power Electron*. May 1997;12(3): 528-536. DOI: 10.1109/63.575680

[22] Bazzi A.M, Dominguez-Garcia A, Krein PT. Markov reliability modeling for induction motor drives under field-oriented control. *IEEE Trans Power Electron*. 2012;27(2): 534-546. DOI: 10.1109/TPEL.2011.2168543

[23] Zeraoulia M, Benbouzid MEH, Diallo D. Electric motor drive selection issues for HEV propulsion systems: a comparative study. *IEEE Trans Vehicular Technol*. Nov. 2006;55(6):1756-1764. DOI: 10.1109/TVT.2006.878719

[24] Morimoto S, Sanada M, Takeda Y. Wide-speed operation of interior permanent magnet synchronous motors with high-performance current regulator. *IEEE Trans Indus Applic*. Jul/Aug 1994;30(4):920-926. DOI: 10.1109/28.297908

[25] Halder S, Srivastava SP, Agarwal P. Flux weakening control algorithm with MTPA control of PMSM drive. In: 2014 IEEE 6th India International Conference on Power Electronics (IICPE); Dec 8-10; Kurukshetra, India. 2014. p. 1-5. DOI: 10.1109/IICPE.2014.7115810

[26] Krishnan R. *Permanent Magnet Synchronous and Brushless DC Motor Drives*. Boca Raton, USA: CRC Press; 2009. 611 p. DOI: 10.1201/9781420014235-P3

Chapter 2

Modeling and Control Strategy for Hybrid Electrical Vehicle

Vu Trieu Minh, Alina Sivitski, Mart Tamre and Igor Penkov

Additional information is available at the end of the chapter

http://dx.doi.org/10.5772/61415

Abstract

This chapter reviews the developments and configurations of hybrid electrical vehicles. A classic model for a parallel hybrid electrical vehicle is chosen and modeled. Model predictive controllers and simulations for this vehicle model are applied to control the vehicle speed and power to check the ability of the system to handle the transitional period for the automatic clutch engagement from the electrical driving to the internal combustion engine (ICE) driving. The chapter produces potential model predictive control considerations to achieve the optimal real-time control actions subject to the vehicle physical constraints. The new system can be applied for electronic control units in real hybrid vehicle powertrains.

Keywords: Model Predictive Control, Automatic Clutch Engagement, Hybrid Vehicle, Powertrain System

1. Introduction

Transportation accounts for more than 25% of the world energy demand and consumes more than 60% of the oil used each year. Transportation is completely relied by almost 95% on the petroleum products. Lack of fossil fuel supplies and the negative greenhouse effect on the environment have motivated automotive engineers to fabricate new car generations to cope the fuel consumption and emissions issues. Hybrid vehicles are now a new product of the automobile industry. Such vehicles are becoming more impressive as they are very efficient and reduce pollution. Hybrid vehicles offer a great improvement on the air quality emissions for vehicles that are empowered by gasoline.

Hybrid electric vehicles (HEVs) are the type of hybrid vehicles and electric vehicles which combine the best features of internal combustion engines (ICEs) and electric motors (EMs).

Hybrid vehicles can perform with much less emission and can save 50% less fuel than the other new conventional vehicles in the same class. Modern hybrid vehicles can now recharge the electrical power from the regenerative braking process and store the electrical energy from the electrical plug-in at outdoor parking locations using renewable energy resources. Hybrid electrical vehicles can completely remove the idle emissions by shutting down the combustion engines and restart them when running.

There are many advantage reasons for using the combination of ICEs and EMs over ICEs alone: EMs can stop completely during the ICEs idle period. EMs can use much less energy than ICEs in low speed (less than 50 km/h) where most of vehicles have to operate in cities. While ICE vehicles can operate better only on the highways with high speed (above 50 km/h) since the ICEs achieve higher level of power-to weight ratio. Therefore, the main advantage of ICEs is on the higher power that can provide to vehicles. Then, the combination of ICEs and EMs can maintain the optimal operation in both low speed on busy roads and high speed on highways. The ICEs can automatically turn on and recharge the batteries for EMs when they get low.

A distinction of HEVs can be divided into two main types: the serial type and parallel types. In the serial type, the ICEs are not mechanically connected to the vehicle powertrain, but they are used only to run the electrical generator to supply the electrical power to batteries for EMs. The EMs now can use the electrical energy from their batteries to propel the vehicle. In the parallel type, both ICE and EM power sources are independently operated; therefore, they can both individually or commonly propel the vehicles. The most general configuration is that the ICE and EM are connected by an automatically controlled clutch. For the EM driving in low speeds, the clutch between the ICE and the EM is open and only the EM runs the vehicle. In high speeds, this clutch is closed and the ICE is started and propelled the vehicle while the EM is stopped off. At very high speeds or at critical heavy loads, this EM can be automatically activated on to support the ICE to drive the vehicle. To conclude, serial hybrids are less efficient and are more suitable for short-distance travel before exhausting their batteries. However parallel hybrids fuel consumption is greater than serial hybrids; gas/diesel engine has dominating role in parallel hybrids and thus battery pack can have less capacity. More information about parallel hybrids in the literature can be referred in [1, 2], and [3] references.

Initial approaches to switch the transition between EM and ICE power sources were drawn on heuristic information on the characteristics of ICEs vs EMs. In reference [4], rule-based was used and in reference [5], fuzzy logic was used as typical controlled schemes. A set of rules was applied to separate the requirements between the two power sources. These control plans were introduced in the early hybrid implementations. However [4] and [5] did not fully investigate the fuel use optimization. The dynamic engagements of the clutch between the ICE and the EM must be controlled as smooth as possible at the right time of engagement. For the safety and comfort reasons, the clutch engagement and the torque transmission are not permitted to cause any unacceptable acceleration or exceeded jerk. Several other control strategies for smothering the clutches engagement have been attempted including the back stepping control in reference [6], the optimal control in the reference [7], and the model predictive control in reference [8] to achieve faster and smoother clutches engagement between ICE and EM. In our research, a new predictive model for controlling the speeds of ICEs and

EMs is developed and applied to the engagement of the clutches, which can help to enhance the driving comfort and reduce the jerk on the parallel HEVs.

The main motivation of using the model predictive control (MPC) scheme in this chapter is the ability of the MPC to determine the optimization actions online in both linear and nonlinear systems. Model predictive control involves new mathematic algorithms that calculate an infinite sequence of input and output variables in order to optimize the future behavior of the systems. To date, MPC strategy can be found in many application areas including aerospace, automotive and petrochemical industries as referred in reference [9]. One of the advantages of MPC is its ability to cope with constraints from the open-loop optimal control problems. Achieving solution from the general constrained nonlinear models within an infinite prediction sequence becomes impractical because the numerical means to solve these problems can be only applied for a finite horizon length in order to obtain a real-time numerical solution. Therefore, only a finite moving horizon regulator can be considered in which the optimal problem is performed over a finite prediction sequence, and the prediction cost after the end of the horizon is estimated from a terminal penalty as shown in reference [10]. For ensuring the stability within a finite prediction length, most of the nonlinear model predictive control (NMPC) strategies have used a terminal constraints region at the end of the prediction length as indicated in references [11, 12], and [13]. Another NMPC method using a terminal constraints region named quasi-infinite horizon that guarantees the asymptotic closed-loop stability with the input constraints was developed in reference [14]. But then the nonlinear systems have both input constraints and output constraints, many difficulties will arise to satisfy the output constraints because of constraints imposed already in inputs. Therefore, another NMPC method without a terminal constraint region is developed in reference [15] using the softened output constraints. Further investigation on MPC such as robust model predictive control (RMPC) that guarantees stability in the existence of the model uncertainty using linear matrix inequalities (LMIs) subject to both input and output saturated constraints was developed in reference [16]. In this RMPC,the controller will soften the output constraints as penalty terms and add into its objective function. These terms will keep the output violation at low values until the constrained solution is returned.

If there are too many input and output constraints, the controller may not be able to provide the desired outputs because the MPC regulator is designed for always on-line implementation, any infeasible solution of the optimization problem cannot be allowed. In order to assure the system stability, the conventional MPC strategies usually want to delete or ignore some output constraints from its objective function. Deletion of some output constraints will make the system looser and then the probability that MPC can achieve an optimal solution will increase. Similarly, the robustness of MPC can also increase if some output setpoints can be relaxed and a new MPC scheme, which turns the constrained setpoints into constrained regions was developed in reference [17]. In our research, the robustness of MPC schemes is also examined with both deleting and changing the output constrained setpoints into constrained regions. In this case, the outputs violation can be regulated by changing the penalty values of the weighting items.

Coming next, we briefly present the characteristics of some typical HEV configurations. After that, one typical HEV configuration will be selected for developing its dynamic equations. Then, several MPC scheme are built to control the speeds of the ICEs and the EMs for each part of this HEV. Simulation and calculated examples are also provided after each section to illustrate our main ideas for each section. Finally, some conclusions are drawn and some suggestions for the future research are discussed.

2. HEVs configurations and reviews

The first concept of HEVs was initiated by Ferdinand Porsche at Lohner Coach Factory when he designed the "Mixte", a series hybrid vehicle based on his earlier electrical vehicle. This hybrid vehicle was installed with a gasoline ICE connected to an electrical generator and an EM propels the vehicle with a small battery for more reliability. This serial HEV concept grew until the late of 1960s. In 1972, Viktor Wouk developed his first prototype for parallel hybrid powertrain in General Motors namely "Godfather of the Hybrid". This turning point ignited the very fast growing progress in HEVs as we can see nowadays and branched in many different aspects such as regenerative braking issues, fuel consumption, emission, battery problems and so on. Worldwide sales of hybrid vehicles reached more than four (4) million units by December 2010 and sold over 80 countries and regions, led by the United States with more 2.5 million units. Regenerative braking, shutting down the engine at idle have reduced the fuel consumption and emissions. The main problem for HEVs now is relying on the cost and the weight of the large batteries and the starter/generator motors.

The recent technology of HEVs is the development of the diesel HEVs since diesel prices are cheaper than gasoline while diesels produce more energy, suffer less wear while operating at higher efficiency. Diesel ICE produces higher torque and offer longer mileage. Most diesel ICE can use 100% pure biodiesel and they don't need petroleum. Diesel ICEs are 20% to 40% more efficient and produce less carbon-dioxide emissions than gasoline ICEs. Diesels are widely popular in Europe, accounting for more than 50% of the car market there. If the diesel HEVs were in use, higher benefits from this system can be achieved.

2.1. Definitions

Conventional vehicles are propelled by only igniting fossil fuel in an ICE and converting the ignition energy into mechanical rotation and translation. In contrast, HEVs are characterized by using some combination of a primary propulsion unit (PPU) that can be fuel cell (FC) of an ICE or EM that can be either an electrochemical storage system such as a battery or an electrostatic super capacitor [22]. In addition to the above mentioned components, at least one electric motor is necessary in any HEV to help propel the vehicle either fully or partially. This combination of electric and fossil fuel energy, supervised with a high level controller called energy management strategy (EMS), can improve the performance of the vehicle from a fuel consumption and emission point of view. Comparing HEVs with conventional vehicles shows

http://dx.doi.org/10.5772/61415

that the former is more fuel efficient due to the engine operation optimization and the possibility of recovering the kinetic energy during braking [23].

EMs play the role of optimizing the efficiency of the ICE as well as energy recovery during braking. It can also use the excess power of the engine to charge the battery if the power demands on the final drive is lower than the power converted by the engine. Another role can be to assist the ICE in the cases that the ICE alone cannot fulfill the driver demands when the ICE is overloaded in some emergent cases.

There are basically four main advantages in hybridizing a conventional vehicle as follows:

1. Engine downsizing: When we use both EM and ICE in one single vehicle, it is possible to size the engine for the mean instead of the peak power demand. In such a way, the electrical buffer (EB) can compensate the lack of power in high power demand periods of driving via the EM. Hence, having a smaller ICE, the vehicle can be driven more efficiently in normal driving compared to having a larger ICE.

2. Regenerative braking possibility: The energy that is dissipated in conventional vehicles during braking can be regenerated and stored in the electric buffer using the electrical machine in its generator mode.

3. Pure electric drive: Having an EM together with the ICE has also advantages of letting the ICE shut down during the vehicle's low speeds in such a way that the controller shuts the ICE down and makes the EM propel the vehicle at low speed. This can also reduce both fuel consumption and emissions.

4. Improve control of the ICE: Since in HEVs, the propulsion power demand is a mix of power from the PPU and the EB, control of the ICE operating point, i.e., engine torque and speed, can be carried out with higher degree of freedom compared to a conventional vehicle and depending on which type of HEV we have. In addition, having an EM in the propulsion system responding to quick changes in propulsion power demand due to its smaller time constant compared to that of the ICE, gives the possibility to avoid transient ICE utilization.

Although HEVs have many advantages, they have some limitations as well. The first issue is the increased cost due to the presence of EMs, energy storage system, electrical converters, and so on. Safety issues due to existence of high voltage electricity and electromagnetic interference due to high frequency switching are also other problems to HEVs.

2.2. HEV configurations

As mentioned above, any HEV contains at least two sources of energy. These two sources can be coupled together in several different combinations according to the application. The most famous configurations are called series, parallel, series-parallel and heavy HEV [23]. The latter is though not a separate category but either a series or a parallel hybrid that is used in heavy delivery vehicles can run on gasoline or diesel. The following is a brief description of each configuration.

Series HEVs:

Series powertrain architecture is used in large vehicles such as locomotives and heavy duty trucks and not in passenger vehicles according to efficiency problems due to their component inefficiencies [24]. In this configuration, the ICE is not directly connected to the final transmission but via two electric machines. In such architecture, the engine mechanical output is first converted into electric energy using an electric generator and then this electric energy can either charge the battery when needed or bypass the buffer and propel the vehicle via a separate electric motor or a combination of these as is shown in Figure 1. Regenerative braking is also possible using the traction motor in generator mode and storing the electric energy into the buffer. So a series HEV needs three machines: one engine, one electric generator, and one electric traction motor.

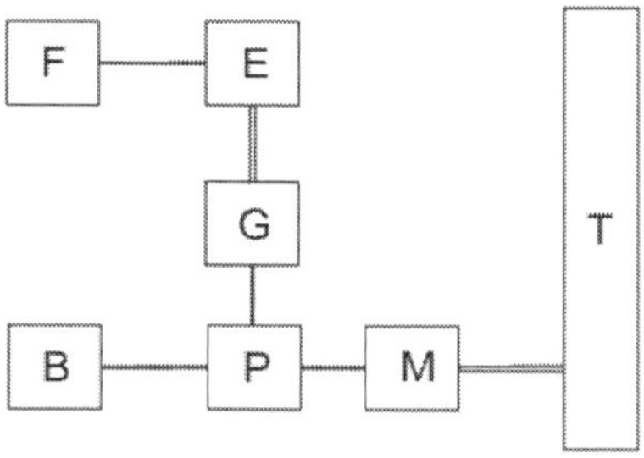

B: Battery; **E**: ICE; **F**: Fuel tank; **G**: Generator; **M**: Motor; **P**: Power converter; **T**: Transmission (including brakes, clutches and gears)

Figure 1. Series HEV Configuration.

Parallel HEVs:

Since the model that is used in this study is a parallel hybrid, it is described here in more details. In parallel hybrid electric vehicles the propulsion energy is delivered to the final drive by the EM and the ICE in two separate parallel paths. Though the ICE is the primary propulsion unit, the EM will be considered as an assistant to the PPU. As shown in Figure 2, both the ICE and the EM are coupled to the final transmission via mechanical links in parallel HEVs. Also, it is worth mentioning that each energy flow path can have a separate clutch to engage/disengage the final transmission from either the ICE and/or the EM. Hence, six (6) different operating modes are possible for a parallel HEV as follows:

1. Motor assist mode: Power from both the ICE and the EM are mixed together to provide the final drive power request. This can be the case in peak power demands. See Figure 3**A**.

2. Regenerative breaking mode: In this mode, the braking energy can be converted into electric energy, using the traction motor working in generator mode, and be stored into the electric buffer via the lower path. See Figure 3**B**.

3. Power split mode: This is the case when the ICE delivers more power than the power demand on final drive and the electric buffer is somehow empty (better say, its charge is close to its lower limit). So, ICE power is split in such a way that it drives the vehicle and the excess power charges the EB at the same time. In this mode, the electric machine works in generator mode. See Figure 3**C**.

4. Motor alone mode: In this mode, the ICE is turned off and the vehicle is fully powered by the EM. In this case, no fuel is used and hence there is no emission. This situation can happen mostly in idling or low power demand driving cases. See Figure 3**D**.

5. ICE alone mode: In ICE alone mode, the vehicle is propelled by the ICE and the EM is switched or via an electric circuit. See Figure 3**E**.

6. Stationary charging mode: In this mode, the vehicle is at standstill, all machineries are off and the vehicle is plugged into the power outlet.

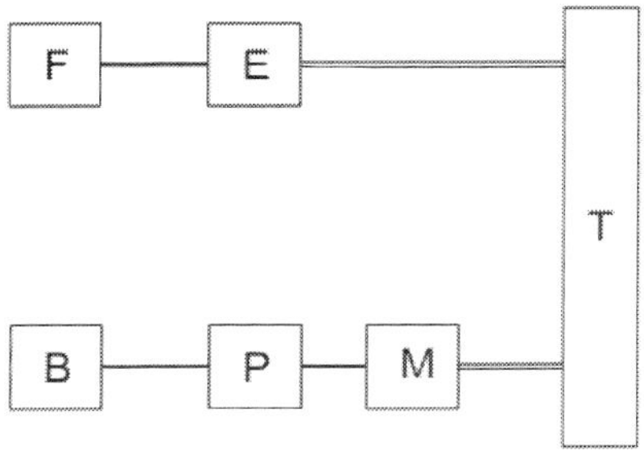

Figure 2. Parallel HEV configuration.

As can be seen in Figure 2, one single electric machine can be used in parallel HEVs, working either as a motor or as an electric generator. This results in a decrease in expenses as well as in total vehicle mass that can prevent an excess in fuel consumption due to the vehicle weight. Different operating modes of a parallel HEV are shown in Figure 3. Note that the arrowheads show the direction of power flow in each operating mode of the vehicle powertrains.

Series-parallel HEVs:

As shown in Figure 4, a third configuration of HEVs can be realized by just adding another power flow path to the parallel architecture having the features of both series and parallel HEV. In this configuration, an extra electric machine is added to the system via a mechanical and an electrical link so that all advantages of series and parallel hybrids can be kept together in one single vehicle. The disadvantage of series-parallel HEVs is though they have complex and costly structure, they are still preferred to be employed in some special cases.

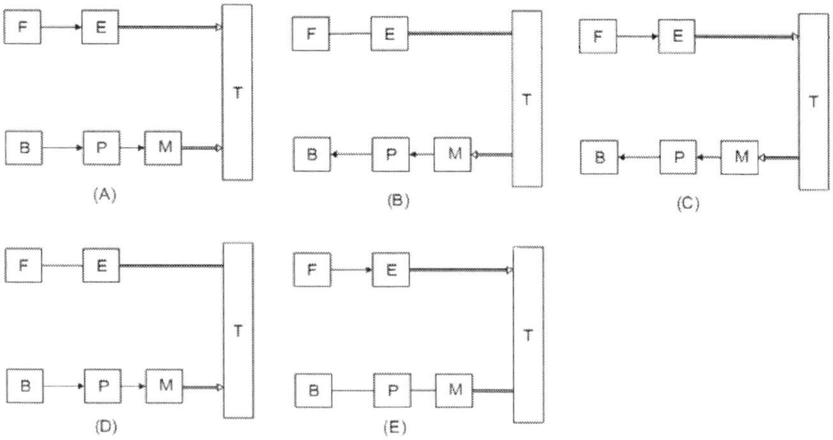

Figure 3. Power flow in a parallel HEV.

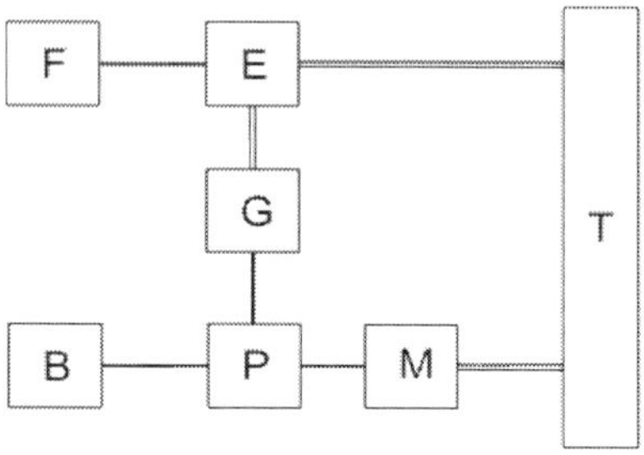

Figure 4. Series-parallel HEV.

2.3. Parallel HEVs reviews

There are four (4) different types of parallel hybrid vehicles: micro-hybrid, mild-hybrid, medium-hybrid, and full-hybrids. The differences are from their functionality relating to the electrical power and the input voltage level (Figure 5). Our research focuses only on the hybrid vehicle with light electrical power supplier up to approximately 10 kW. With these light electric power, some relatively high benefits can be obtained.

The input voltage for a mild-hybrid is usually applied at 36 V, but for this lightest electrical power, such a higher voltage level will not be needed and the 12 V will be a better solution. Due to the ability of some power assist for the regenerative braking, this configuration is categorized as mild-hybrid. The voltage level of medium-hybrid is 144 V while the voltage level for full-hybrid is 300 V. In this research, a full-hybrid vehicle is developed with two EMs.

One main EM uses the synchronous DC 300 V and the other is a synchronous starter/generator DC 48 V.

A conventional vehicle powertrain consists of one electrical starter connecting to one electrical generator. The electrical starter can convert the electrical energy into mechanical torque and is used to rotate the engine. The electrical generator is basically differing, it connects to the crankshaft and runs as an electrical supplier, transforming the ICE's mechanical torque into the electrical energy and stores this power into the vehicle batteries. The integrated starter generator (ISG) is a new development and becomes the only one device that will combine both above functions. For generating mode, it will charge the vehicle batteries and for starting mode, it will discharge the vehicle batteries in order to start or assist the ICE.

There are several possible positions for the ISG to be assembled. Three (3) main typically configurations are, the crankshaft-mounted ISG (C-ISG), the belt driven ISG (B-ISG), and the drivetrain mounted ISG (D-ISG) as shown from Figure 6 to Figure 8. The C- and D-ISG are usually used for the larger electrical power levels than the B-ISG configuration. Reasons for these differences are the limited power suppliers that can be transferred without slip through the belt of the B-ISG configuration. In the case of D-ISG configuration, the clutch is installed between the ICE and the ISG, so that, the hybrid vehicle can operate from either the ISG or on the combination of both power supplier sources. Rationally, if the vehicle can be able to operate completely on the ISG, the electrical power supplier must be large adequate.

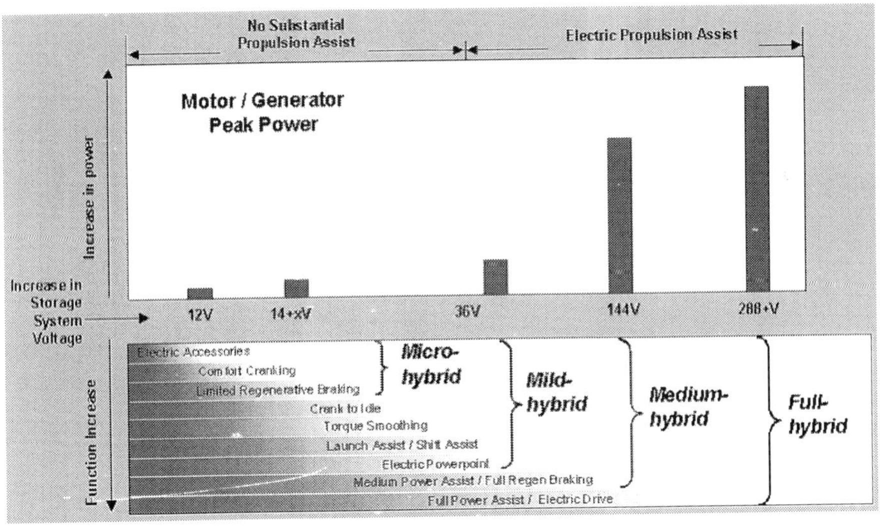

Figure 5. Different classes of hybridization [18].

One clear advantage of the D-ISG configuration regarding to other variants is that during deceleration the drag torque of the ICE can be removed by opening the clutch, letting more

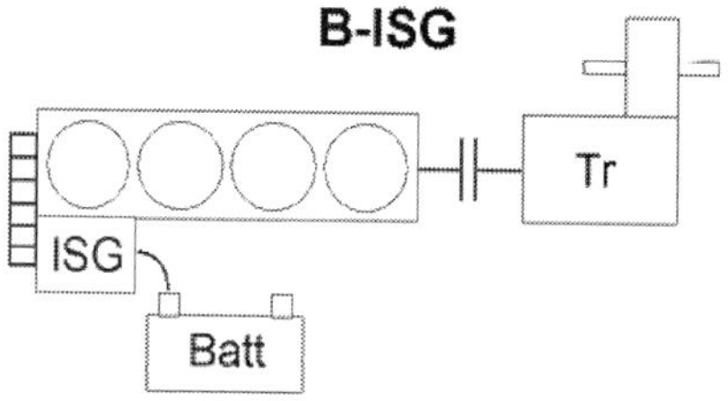

Figure 6. Schematic representation of a B-ISG.

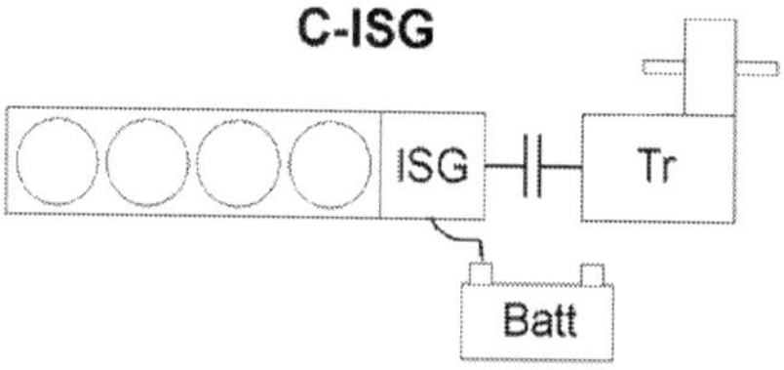

Figure 7. Schematic representation of a C-ISG.

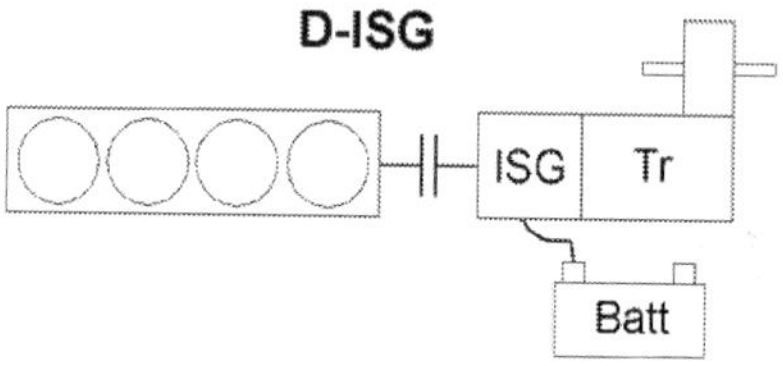

Figure 8. Schematic representation of a D-ISG.

energy available for regenerative braking processes. However, because of the limited charging acceptance of the vehicle batteries, the full ability of regenerative braking process will be never employed unless the deceleration is very slow. Additionally, purchasing a B-ISG system is less cost, which will provide some important advantages over the use of the C-ISG and the D-ISG.

From the above overview of HEV configurations, we will select a typical parallel HEV configuration and start the full investigation with this new HEV model in the next section.

3. HEV configuration and modeling

A typical update HEV for this research is shown in Figure 9. It is from a very common parallel HEV concept system designed by Daimler Chrysler, namely the P12-Configuration appeared at the 2004 Detroit Motor Show. The model concept includes one (1) conventional ICE and two (2) EM1 and EM2. An automatically controllable conventional friction clutch separates the model drivetrain into two (2) parts: Part 1: ICE with EM1, and Part 2: EM2 and the rest of the powertrain transmission. The EM1 operates as an electrical starter and a supported electrical generator. No torque converter is installed in this configuration. The driven wheel of this model is the rear wheel, which is connected with a conventional automated transferred gearbox and a conventional differential gearbox.

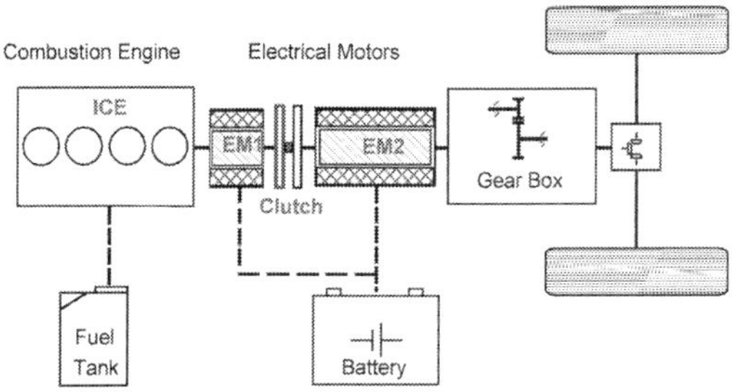

Figure 9. Configuration of the parallel hybrid powertrain.

At the only electrical driving for low speeds (below 50 km/h), the clutch between EM1 and EM2 is opened and the vehicle runs in a series hybrid configuration. For this operating mode, only second motor, EM2, drives the vehicle.

Transmission from the series mode to the parallel mode will be occurred at high speed (above 50 km/h) by closing the friction clutch between EM1 and EM2. EM1 will start the ICE for propelling the vehicle. After the ICE already started, the EM1 will turn off. EM2 will turn on if necessary as an electrical generator to charge the vehicle batteries.

In extreme heavy loads and from the control of the driver, both EM1 and EM2 can be simultaneously activated to assist the ICE to run the vehicle in critical driving conditions.

Our research concentrates on the development of reliable control strategies that can control the speeds of these two drivetrain parts and then, to synchronize these parts with a friction clutch to achieve the fast and smooth clutch engagement with reduction of driveline jerk and improvement of the driving comfort.

Figure 10 shows a simplified dynamic model for this hybrid transmission drivetrain. As referred in the references [19] and [20], the first part of the transmission powertrain can be approximately estimated by a lumped inertia of J_1 where the main shaft holding the ICE, EM1 and one friction clutch plate is modeled as a rigid frame. The second part of this transmission powertrain is also modeled by two inertias of J_2 and J_3 connected by a mechanical and damping springs. The conventional transmission gearbox and the differential gearbox are also modeled as a simple transforming torque with a variable term i depending on the selected gear transmission ratios.

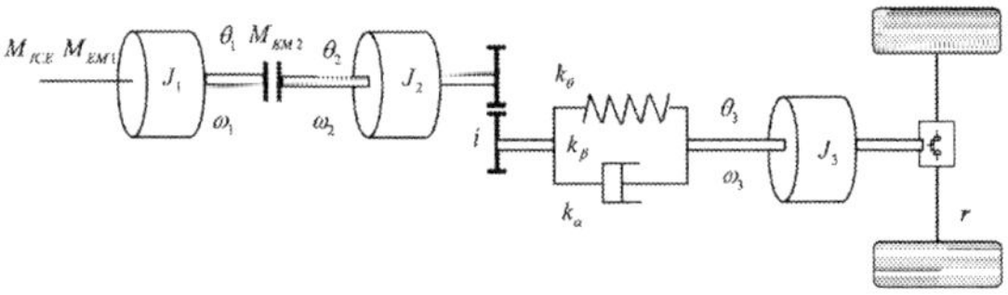

Figure 10. Simplified drivetrain structure.

where J_1 is the lumped inertia of ICE and EM1; J_2 is the inertia of EM2, and J_3 is the lumped inertia of the rest of the transmission powertrain. The transmission powertrain can now be considered as a connection of a spring with the torque stiffness k_θ, the velocity damping coefficient k_β, and the acceleration damping coefficient k_α. The torques are M_{ICE}, M_{EM1}, and M_{EM2} generated from ICE, EM1, and EM2, respectively. Angular angles from shaft1, shaft2, and shaft3 are θ_1, θ_2, and θ_3. Angular velocities ω_1, ω_2, and ω_3 are measured from shaft1, shaft2, and shaft3 respectively. r is the vehicle wheel rolling radius.

There are three separated influences to the hybrid vehicle resistances by the air drag; the rolling friction resistance and the combined mechanical torque losses in the transmission gearbox; the differential gearbox and the shaft bearings due to friction. They can be approximately estimated from the hybrid vehicle velocity resistance torque, M_v, as:

$$M_v = \left(\frac{\rho}{2} c_w A (r\omega_3)^2 + f_r mg \right) r + a_0 + a_1\omega_3 + a_2\omega_3^2 \tag{1}$$

where drag coefficient; A: vehicle frontal area; ρ: air density; c_w; r: wheel dynamic radius; m: vehicle mass; f_r: resistance coefficient; a_i: polynomial coefficients; and g: natural gravity. The effect of the road dynamics and the road inclination can be considered as the additional disturbances to the system. These additional disturbances can cause some extra accelerating or decelerating torques at the rolling resistances. However, the changes of these effects and the changes of the vehicle load/mass during the vehicle running are not considered in our chapter.

At low speed (less 50 km/h), the powertrain torque model is formulated without the contribution of the exponential term of ω_3^2 to M_v. And the linearization of the resistance torque M_v in the above equation (1) is now calculated as:

$$M_v = M_{v0} + k_v\omega_3 \tag{2}$$

where M_{v0} is the initial constant value for the air drag and k_v is the linear air drag coefficient. Because of the differences in gearbox ratios, the coefficient constant M_{v0} and k_v can be varied for each gearbox ratio.

The following torque equations are formulated for the first powertrain part:

$$M_{1_o} = J_1\dot{\omega}_1 \tag{3}$$

The driving torque is computed as:

$$M_{1_o} = M_{ICE} + M_{M1} - M_C \tag{4}$$

The friction torque M_C transmitted by the clutch can be divided into two engagement modes:

The locked mode when the friction torque (M_C) exceeds the static friction capacity ($M_{f\max}^{Static}$):

$$M_C = \frac{2}{3} r_C F_{NC} \mu_S \text{ when } \left(M_C = M_{f\max}^{Static}\right) \tag{5}$$

where F_{NC}: normal force exerted on the clutch; r_C: clutch corresponding radius; and μ_S: clutch static friction coefficient.

And the slipping mode:

$$M_C = r_C F_{NC} sign(\omega_1 - \omega_2)\mu_K \text{ when } (M_C < M_{f\max}^{Static})$$

where μ_K: clutch slipping kinetic friction coefficient.

The following torque equations are applied for the second part:

$$\begin{aligned} & M_{2_o} = k_\theta \theta_2 + \frac{k_\theta}{i}\theta_3 + k_v \omega_3 \qquad \text{a} \\ & \text{or} \\ & M_{2_o} = J_2 \dot{\omega}_2 i + J_3 \dot{\omega}_3 + k_v \omega_3 \qquad \text{b} \end{aligned} \tag{6}$$

And

$$\dot{M}_{2_o} = J_2 \ddot{\omega}_2 i + k_\alpha \left(\frac{\dot{\omega}_2}{i} - \dot{\omega}_3\right) + k_\beta \left(\frac{\omega_2}{i} - \omega_3\right) \tag{7}$$

The torque M_{2_o} is computed:

$$M_{2o} = \left(M_{EM2} + M_c\right)\eta i - M_{v0} \tag{8}$$

with η: efficiency of gearbox and differential.

The above torque equations can be transformed to the following dynamic equations:

$$\dot{\theta}_1 = \omega_1 \tag{9}$$

$$\dot{\omega}_1 = -\frac{k_{\beta 1}\omega_1}{J_1} + \frac{M_{ICE}}{J_1} + \frac{M_{M1}}{J_1} + \frac{-M_C}{J_1} \tag{10}$$

where $k_{\beta 1}$ is friction coefficient in shaft 1.

$$\dot{\theta}_2 = \omega_2 \tag{11}$$

$$\dot{\omega}_2 = -\frac{k_{\beta 2}\omega_3}{J_2 i} - \frac{J_3\dot{\omega}_3}{J_2 i} - \frac{\eta M_{M2}}{J_2} + \frac{\eta M_C}{J_2} - \frac{M_{v0}}{J_2 i} \tag{12}$$

where $k_{\beta 2}$ is friction coefficient in shaft 2.

$$\dot{\theta}_3 = \omega_3 \tag{13}$$

$$\dot{\omega}_3 = \frac{k_{\beta 3}\omega_3}{J_3} + M_{v0} \tag{14}$$

where $k_{\beta 3}$ is friction coefficient in shaft 3.

$$\ddot{\omega}_3 = \frac{k_{\beta 2}\omega_2}{J_3 i} - \frac{(k_{\beta 2}J_2 i^2 + k_\alpha k_v)\omega_3}{J_2 J_3 i^2} - \left(\frac{k_v + k_\alpha}{J_3} + \frac{k_\alpha}{J_2 i^2}\right)\dot{\omega}_3 + \frac{k_\alpha \eta (M_{M2} + M_C)}{J_2 J_3 i} - \frac{k_\alpha M_{v0}}{J_2 J_3 i^2} \tag{15}$$

Replace the torque generated by a DC motor in the following formula:

$$M_{DC\_MOTOR} = \frac{k_T}{R_I}V - \frac{k_E k_T}{R_I}\omega \tag{16}$$

http://dx.doi.org/10.5772/61415

where M_{DC_MOTOR}: torque by DC Motor; k_T: motor torque constant, $k_T = \frac{M_{Torque}}{I_{Current}}$ (Nm/A); k_E: motor electromotive force (EMF) constant (V/ (rad/s)), $k_E = k_T$; R_I: terminal resistance (Ohm), V: power supply (Volts), and ω: angular velocity (rpm).

Then a new set of vehicle dynamics is installed as:

$$\dot{\theta}_1 = \left[0+ \quad \omega_1 + \quad 0+ \quad 0+ \quad 0+ \quad 0+ \quad 0\right] + \left[0+ \quad 0+ \quad 0+ \quad 0+ \quad 0\right] \tag{17}$$

$$\dot{\omega}_1 = \left[0+ \quad \frac{-\left(k_{\beta 1} + \frac{k_{E1}k_{T1}}{R_{I1}}\right)\omega_1}{J_1} + \quad 0+ \quad 0+ \quad 0+ \quad 0+ \quad 0\right] + \left[\frac{M_{ICE}}{J_1} + \quad \frac{k_{T1}V_1}{R_{I1}J_1} + \quad 0+ \quad \frac{-M_C}{J_1} + \quad 0\right] \tag{18}$$

$$\dot{\theta}_2 = \left[0+ \quad 0+ \quad 0+ \quad \omega_2 + \quad 0+ \quad 0+ \quad 0\right] + \left[0+ \quad 0+ \quad 0+ \quad 0+ \quad 0+\right] \tag{19}$$

$$\begin{aligned}\dot{\omega}_2 = &\left[0+ \quad 0+ \quad 0+ \quad 0+ \quad 0+ \quad \frac{-\left(k_{\beta 2} + \frac{k_{E2}k_{T2}}{R_{I2}}\right)\omega_3}{J_2 i} + \quad \frac{-J_3\dot{\omega}_3}{J_2 i}\right] \\ &+\left[0+ \quad 0+ \quad \frac{-\eta k_{T2}V_2}{R_{I2}J_2} + \quad \frac{\eta M_C}{J_2} + \quad \frac{-M_{v0}}{J_2 i}\right]\end{aligned} \tag{20}$$

$$\dot{\theta}_3 = \left[0+ \quad 0+ \quad 0+ \quad 0+ \quad 0+ \quad \omega_3 + \quad 0\right] + \left[0+ \quad 0+ \quad 0+ \quad 0+ \quad 0\right] \tag{21}$$

$$\dot{\omega}_3 = \left[0+ \quad 0+ \quad 0+ \quad 0+ \quad 0+ \quad \frac{k_{\beta 3}\omega_3}{J_3} + \quad 0\right] + \left[0+ \quad 0+ \quad 0+ \quad 0+ \quad M_{v0}\right] \tag{22}$$

$$\begin{aligned}\ddot{\omega}_3 = &\left[0+ \quad 0+ \quad 0+ \quad \frac{-\left(k_{\beta 2} + \frac{k_{E2}k_{T2}}{R_{I2}}\right)\omega_2}{J_3 i} + \quad 0+ \quad \frac{-\left(k_{\beta 2}J_2 i^2 + k_\alpha k_v\right)\omega_3}{J_2 J_3 i^2} + \quad -\left(\frac{k_v + k_\alpha}{J_3} + \frac{k_\alpha}{J_2 i^2}\right)\dot{\omega}_3\right] \\ &+\left[0+ \quad 0+ \quad \frac{-k_\alpha \eta k_{T2}V_2}{R_{I2}J_2J_3 i} + \quad \frac{k_\alpha \eta M_C}{J_2J_3 i} + \quad \frac{-k_\alpha M_{v0}}{J_2J_3 i^2}\right]\end{aligned} \tag{23}$$

If we define the state variables as $x_0=[\theta_1 \ \ \omega_1 \ \ \theta_2 \ \ \omega_2 \ \ \theta_2 \ \ \omega_3 \ \ \dot{\omega}_3]'$ for the positions, angular velocities, and acceleration on vehicle shafts 1, 2, 3 respectively, and the input variables as $u_0=[M_{ICE} \ \ V_1 \ \ V_2 \ \ M_C \ \ M_{v0}]'$ for the torque for ICE, voltage for EM1; voltage for EM2, torque on clutch, and the initial air-drag torque load, a space state form of the vehicle dynamics is set up:

$$\dot{x}_0 = \begin{bmatrix} 0 & 1 & 0 & 0 & 0 & 0 & 0 \\ 0 & \dfrac{-\left(k_{\beta 1} + \dfrac{k_{E1}k_{T1}}{R_1}\right)}{J_1} & 0 & 0 & 0 & 0 & 0 \\ 0 & 0 & 0 & 1 & 0 & 0 & 0 \\ 0 & 0 & 0 & 0 & 0 & \dfrac{-\left(k_{\beta 2} + \dfrac{k_{E2}k_{T2}}{R_2}\right)}{J_2 i} & \dfrac{-J_3\dot{\omega}_3}{J_2 i} \\ 0 & 0 & 0 & 0 & 0 & 1 & 0 \\ 0 & 0 & 0 & 0 & 0 & \dfrac{k_{\beta 3}\omega_3}{J_3} & 0 \\ 0 & 0 & 0 & \dfrac{-\left(k_{\beta 2} + \dfrac{k_{E2}k_{T2}}{R_2}\right)}{J_3 i} & 0 & \dfrac{-\left(k_{\beta 2}J_2 i^2 + k_\alpha k_v\right)}{J_2 J_3 i^2} & -\left(\dfrac{k_v + k_\alpha}{J_3} + \dfrac{k_\alpha}{J_2 i^2}\right) \end{bmatrix} x_0 + \begin{bmatrix} 0 & 0 & 0 & 0 & 0 \\ \dfrac{1}{J_1} & \dfrac{k_{T1}}{R_1 J_1} & 0 & \dfrac{-1}{J_1} & 0 \\ 0 & 0 & 0 & 0 & 0 \\ 0 & 0 & \dfrac{\eta k_{T2}}{R_2 J_2} & \dfrac{\eta}{J_2} & \dfrac{-1}{J_2 i} \\ 0 & 0 & 0 & 0 & 0 \\ 0 & 0 & 0 & 0 & 1 \\ 0 & 0 & \dfrac{k_\alpha \eta k_{T2}}{R_2 J_2 J_3 i} & \dfrac{k_\alpha \eta}{J_2 J_3 i} & \dfrac{-k_\alpha}{J_2 J_3 i^2} \end{bmatrix} u_0 \qquad (24)$$

The new dynamic modeling in (24) allows having a deep viewing on the acceleration $\dot{\omega}_3$ and jerk $\ddot{\omega}_3$ of HEVs. This is also one of the mail contributions of this study.

When the vehicle travels in low speed (below 50 km/h), only EM2 is running. Then, the inputs: $M_{ICE}=0$, $V_1=0$, $M_C=0$ and the state variables: $\theta_1=0$, $\omega_1=0$, the system becomes:

$$\begin{cases} \dot{x}_p = A_p x_p + B_p u_p \\ y_p = C_p x_p + D_p u_p \end{cases} \text{with}$$

$$A_p = \begin{bmatrix} 0 & 1 & 0 & 0 \\ 0 & -\dfrac{k_{\beta 2} + \dfrac{k_{E2}k_{T2}}{R_{I2}}}{J_2} & 0 & 0 \\ 0 & 0 & 0 & 1 \\ 0 & 0 & 0 & -\dfrac{(k_{\beta 3} - k_v)}{J_3} \end{bmatrix}, B_p = \begin{bmatrix} 0 & 0 \\ \dfrac{k_{T2}}{R_{I2}J_2} & 0 \\ 0 & 0 \\ 0 & -1 \end{bmatrix} \tag{25}$$

$$C_p = \begin{bmatrix} 0 & 0 & 0 & 1 \\ k_\theta & 0 & \dfrac{k_\theta}{i} & 0 \end{bmatrix}; D_p = \begin{bmatrix} 0 & 0 \\ 0 & 1 \end{bmatrix}$$

where $x_p = [\theta_2\ \omega_2\ \theta_3\ \omega_3]'$, $u_p = [V_2\ \ M_{v0}]'$, the outputs, $y_p = [\omega_3\ \ T_{Torque3}]'$, are the vehicle velocity (measured) ω_3 and the vehicle torque (unmeasured) $T_{Torque3}$ generated on shaft 3. In this case, the torsional rigidity (Torque/angle): $k_\theta = \frac{M_{Torque}}{\varphi} = \frac{GJ}{l}$, where φ is the angle of twist $\varphi = \theta_2 - \frac{\theta_3}{i}$ (rad). G is the shear modulus or modulus of rigidity of mild carbon steel, $G = 81500.10^6$ (N/m). l is the length of the shaft where the torque is being applied, in this case, we assumed: $l = 1.5$ (m) for the vehicle drivetrain length. J is the moment of inertia, $J = J_2 + J_3$ (m^4).

At the high speed (more than 50 km/h), EM1 starts ICE while the clutch is still open, the dynamic equations in the first part become:

$$\dot{\theta}_1 = \begin{bmatrix} 0 + & \omega_1 \end{bmatrix} + \begin{bmatrix} 0 + & 0 \end{bmatrix} \tag{26}$$

$$\dot{\omega}_1 = \begin{bmatrix} 0 & -\dfrac{\left(k_{\beta 1} + \dfrac{k_{E1}k_{T1}}{R_{I1}}\right)}{J_1}\omega_1 \end{bmatrix} + \begin{bmatrix} \dfrac{\varsigma k_{T1}}{R_{I1}J_1} V_1 + & \dfrac{1}{J_1} M_{ICE} \end{bmatrix} \tag{27}$$

where ς is a new coefficient added to the EM1 as a compensated load to the starting period.

Then the system becomes:

$$\begin{cases} \dot{x}_e = A_e x_e + B_e u_e \\ y_e = C_e x_e + D_e u_e \end{cases}, \text{ with}$$

$$A_s = \begin{bmatrix} 0 & 1 \\ 0 & -\dfrac{\left(k_{\beta 1} + \dfrac{k_{T1}^2}{R_{I1}}\right)}{J_1} \end{bmatrix}; B_s = \begin{bmatrix} 0 & 0 \\ \dfrac{\varsigma k_{T1}}{R_{I1} J_1} & \dfrac{1}{J_1} \end{bmatrix} \tag{28}$$

$$C_s = \begin{bmatrix} 0 & 1 \\ 0 & \dfrac{\left(k_{\beta 1} + \dfrac{k_{E1} k_{T1}}{R_1}\right)}{J_1} \end{bmatrix}; D_s = \begin{bmatrix} 0 & 0 \\ 0 & 0 \end{bmatrix}$$

where, $x_e = [\theta_1 \quad \omega_1]'$, $u_e = [V_1 \quad M_{ICE}]'$, $y_e = [\omega_1 \quad T_{Torque1}]'$. $T_{Torque1}$ is the output torque (unmeasured) on shaft 1.

Using comprehensive HEV modeling equations from (24) to (28), we can develop MPC controllers to this HEV in the next section.

4. Model predictive controller design

MPC is one of the advanced control theories that have been studied extensively by the research community. MPC provides the optimal solutions for the open-loop manipulated input trajectory that minimizes the difference between the predicted plant behavior and the desired plant behavior. MPC diverges from other control techniques in that the optimal control problem is solved on-line for the current state of the plant, rather than off-line as a feedback control policy. MPC has been broadly used in industries because of its ability to deal with the input and output constraints in the optimal control problems.

The success of model predictive control is greatly depending on the exactness in the open-loop horizon predictions, which in turn rely on the exactness of the plant models. Several issues about MPC still remain open and are of interest to researchers due to the lack of a theoretical basis such as offset-free properties and robustness of MPC toward the environment disturbances. In this research, we study the ability of MPC controllers for the HEVs subject to input and output constraints.

In this research, the duties of MPC are applied to control the output torques generated and transferred among the components and the velocities of each shafts for achieving smoother clutch engagements and higher driving comfort. The RMPC schemes for uncertain systems subject to input and output saturated constraints are referred to the reference in [16]. The predictive control with soften output constraints is referred to the reference in [17] when a new MPC controller with output regions is developed to improve the robustness of the controller for handling input and output constraints and rejecting disturbances. NMPC conditions for stability with soften output constraints are referred to in reference [15]. Development of fault detection for control system using MPC is referred to in reference [21] where the MPC schemes can be reconfigured as the real-time for detecting faults to maintain the error free for the system. Some other newer strategies of HEVs and MPC are referred to in references [25, 26], and [27].

The formula in (24) can now be discretized into the following format:

$$\begin{cases} x_{t+1} = Ax_t + Bu_t \\ y_t = Cx_t + Du_t \end{cases}, \tag{29}$$

where x_t, u_t, and y_t are state variables, inputs and outputs, respectively; A, B, C, and C are static matrices.

The above system is subject to the following input and output constraints:

$$u_t \in \mathcal{U}, \Delta u_t = u_t - u_{t-1} \in \Delta\mathcal{U}, \text{ and } y_t \in \mathcal{Y} \tag{30}$$

The optimal problem for this MPC controller tracking some output setpoints can be presented in the following objective function:

$$\begin{aligned}
&\min_{U \triangleq \{\Delta u_1, \ldots, \Delta u_{1+N_u-1}\}} \left\{ J(U, x(t)) = \sum_{k=0}^{N_y-1} \left[(y_{t+k|t} - r)' Q (y_{t+k|t} - r) + \Delta u'_{t+k} R \Delta u_{t+k} \right] \right\}, \\
&\text{subject to}: u_{t+k} \in \left[u_{\min}, u_{\max} \right], \Delta u_{t+k} \in \left[\Delta u_{\min}, \Delta u_{\max} \right], \text{ for } k = 0, 1, \ldots, N_u - 1 \\
&y_{t+k|t} \in \left[y_{\min}, y_{\max} \right], \text{ for } k = 0, 1, \ldots, N_y - 1 \\
&\Delta u_{t+k} = 0, \text{ for } k = 0, 1, \ldots, N_y - 1 \\
&x_{t|t} = x(t), x_{t+k+1|t} = Ax_{t+k|t} + Bu_{t+k}, u_{t+k|t} - u_{t+k-1|t} \mid \Delta u_{t+k|t}, y_{t+k|t} = Cx_{t+k|t} + Du_{t+k|t}
\end{aligned} \tag{31}$$

is solved at each time t, where $x_{t+k|t}$ denotes the predicted state vector at time $t+k$, obtained by applying the input increment sequence, $U \triangleq \{\Delta u_t, \ldots, \Delta u_{t+N_u-1}\}$, and the new inputs, $u_{t+k|t} = u_{t+k-1|t} + \Delta u_{t+k|t}$, to the model equation (29) starting from the state $x_t = x(t)$. $y_{t+k|t}$ and r are the predicted output variables and the output set-points, respectively. The output set-points can now be reformulated depending on the desired speeds by the driver or $r = r(t)$. The weighting matrices of $Q = Q' \geq 0$ and $R = R' > 0$ are applied for the predicted outputs and the input increments.

For MPC regulator tracking setpoints, the steady-state variables are kept equal to the target set-points when there are no disturbances and constraints. Formula (31) is the one that we apply for the remainder of this chapter to test the ability of MPC to control the HEV velocities.

In this research, we assume that the MPC outputs horizon length is set equal to the inputs control horizon, i.e., $N_u = N_y = N_p$ (equal to the predictive lengths). The quadratic objective junction $J(U, x(t))$ in equation (4.3) is minimized over a vector N_p future prediction inputs starting from the state $x(t)$.

For the MPC with hard constraints, by substituting $x_{t+N_p|t} = A^{N_p}x(t) + \sum_{k=0}^{N_p-1} A^k Bu_{t+N_p-1-k}$, equation (31) can be rewritten as a function of the current state $x(t)$ and the current setpoints $r(t)$:

$$\Psi\left(x(t), r(t)\right) = \frac{1}{2}x'(t)Yx(t) + \min_{U}\left\{\frac{1}{2}U'HU + x'(t)r(t)FU\right\}, \tag{32}$$

subject to the hard combined constraints of $GU \leq W + Ex(t)$, where the column vector $U \triangleq \left[\Delta u_t, \ldots, \Delta u_{t+N_p-1}\right]' \in \Delta\mathcal{U}$ is the MPC optimization vector; $H = H' > 0$, and then, H, F, Y G, W and E are proceeding matrices obtained from Q, R and $x(t)$, $r(t)$ in equation (4.1). Because that we use only the optimizer U, other terms involving Y are normally removed from equation (32). Therefore, the optimal problem in (4.4) will be a purely quadratic formula and depending on only the current state variables $x(t)$, and the current set-points $r(t)$. Implementation of MPC always requires the real-time solution of each quadratic program for each discrete time steps.

In fact, the system may have always both input and output constraints. Difficulties will be arised due to the inability to respond to all output constraints because of the already input constraints. Since MPC is applied for the real-time implementation, any infeasible solution of the optimal control problems cannot be tolerated. Basically if the input constraints are set from the system physical limits and usually considered as the hard/unchanged constraints. At the same time, if the system outputs constraints are the measured velocities and the unmeasured torques which are not so much strictly imposed and can be violated somewhat during the movement of the vehicles. In order to guarantee the system stability if some outputs may violate the constraints, equation (32) can be transformed to some other soften constraints.

$$\begin{gathered}\Psi\left(x(t), r(t)\right) = \frac{1}{2}x'(t)Yx(t) + \min_{U}\left\{\frac{1}{2}U'HU + x'(t)r(t)FU + \frac{1}{2}\varepsilon_i'(t)\Lambda\varepsilon_i(t)\right\}, \\ \text{subject to } \varepsilon_i(t) = \left[\varepsilon_y; \varepsilon_u\right], y_{\min} - \varepsilon_y \leq y_{t+k|t} \leq y_{\max} + \varepsilon_y \text{ and } u_{\min} - \varepsilon_y \leq u_{t+k|t} \leq u_{\max} + \varepsilon_u\end{gathered} \tag{33}$$

The new weighting items, $\varepsilon_i(t)$, are added into the MPC soften objective function: $\Lambda > 0$ (generally some small values) become the weighting factors, $\varepsilon_i(t)$ are represent the violation penalty terms ($\varepsilon_i(t) \geq 0$) for the scheme objective function. These values will keep the output violations at low levels until the constrained solution can be appeared. Further reference of the MPC's subject to these soften state constraints can be read from reference [16].

In order to increase the ability of MPC to get on-line solutions for some critical times, some of the output set-points can be temporally deleted since if some output set-points are omitted, the system will become looser and then, the possibility that MPC can find some solutions will increase. Temporally omitting of some output set-points can also be performed by putting some zeros into the weighting matrix Q in equation (33). The robustness of MPC scheme can be also increased if some output set-points become relaxed into regions rather than in some hard values. If the system output constraints are set into output regions, the MPC scheme will need to change slightly because the set-points $r(t)$ in equation (33) now turn regions. In this formula, output regions are defined by the minimum and maximum limits in a desired output range. The maximum value is the upper limit and the minimum value is the lower limit, $y_{lower} \le y_{t+k \mid t} \le y_{upper}$. From this range, a modified MPC objective function for the output regions is developed as:

$$\min_{U \triangleq \{\Delta u_1, \ldots, \Delta u_{1-N_u-1}\}} \left\{ J(U, x(t)) = \sum_{k=0}^{N_u-1} \left[z'_{t+k|t} Q z_{t+k|t} + \Delta u'_{t+k} R \Delta u_{t+k} \right] \right\},$$
$$\text{where } z_{t+k|t} \ge 0; z_{t+k|t} = y_{t+k|t} - y_{upper} \text{ for } y_{t+k|t} > y_{upper};$$
$$z_{t+k|t} = y_{lower} - y_{t+k|t} \text{ for } y_{t+k|t} < y_{lower}; z_{t+k|t} = 0 \text{ for } y_{lower} \le y_{t+k|t} \le y_{upper} \quad (34)$$

When the system outputs are still laid inside the desired regions, there is no control action being taken, due to none of the control outputs are violated ($z_{t+k \mid t} = 0$). But if some outputs come to violate the desired regions, MPC regulator in equation (34) will activate the objective function, find out optimal inputs action to push the outputs back to the desired regions. Further MPC developments with output constraints deletion and output regions can be referred in reference [17]. We now illustrate the robustness of MPC for soften output constraints and the constraint regions in the following part.

4.1. MPC control application for EM2

The MPC control for EM2 is used for driving the HEVs at the slow speed (less 50 km/h). In this moment the friction clutch is open and both ICE and EM1 are turned off. The dynamic equation of the system in this case is formulated in equation (3.24). The MPC objective function applied in this system is developed in equations (4.3), (4.5), and (4.6). The discrete time interval of all simulations is set at 0.01 sec.

Parameters used for the EM2 are: torsional rigidity, $k_\theta = 1158$; motor constants, $k_{E2} = k_{T2} = 10$; motor inertia, $J_2 = 1$; load inertia, $J_3 = 2$; gear ratio, $i = 2.34$; motor damping, $k_{\beta 2} = 0.5$; load damping, $k_{\beta 3} = 12$; armature resistance, $R_{I2} = 5$.

The input constraints in this example are set as follows: control constraints for the DC input voltage for the vehicle is $\mid V_2 \mid \le 300V; \Delta u(t) \le \inf$, or in this case, there is no need to set the input limit increments on $\Delta u(t)$. The physical output constraints are set for the motor shaft with an allowable shear stress (carbon steel), $\tau = 25$ (MPa or N/mm²). The constraint for the output torque on shaft2 is $|T| = \tau\pi \frac{d^3}{16}$, where $d = 0.05m$ is the output shaft diameter of the motor. Then, the output torque constraint will be $|T_2| \leq 455Nm$.

The MPC parameters will be selected as the horizon prediction length, $N_u = N_y = N_p = 5$, the weighting matrices, $Q = \begin{bmatrix} 10 & 0 \\ 0 & 10 \end{bmatrix}$ and $R = [1]$. Figure 11 shows the MPC performance with the input voltage, the output speed, and torque.

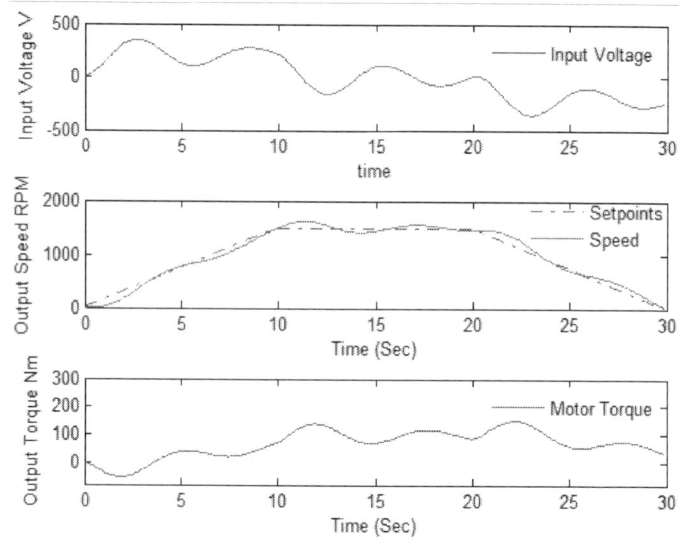

Figure 11. MPC performance with $N_p = 5$ and $Q = 10R$.

Because the MPC performance is highly relying on the selection of the values in weighting matrices Q and R. If we choose that Q is much bigger than R ($Q \gg R$), then, the voltage input control increment $\Delta u(t)$ will become much bigger than the output penalty in $(y(t) - r)$ as indicated in the MPC objective function in (6.3). The controller will drive the vehicle to track the output set-points very fast, but in return, the vehicle will need very much great energy for input torques as illustrated in Figure 12.

In the simulation, the horizon prediction is chosen as $N_u = N_y = N_p = 5$ and the values for weighting matrices are selected as $Q = \begin{bmatrix} 100 & 0 \\ 0 & 100 \end{bmatrix}$ and $R = [1]$:

The MPC horizon length of prediction is highly influenced by its performance. The system will become looser if a longer prediction horizon is selected and the MPC will achieve better performance since the system becomes more flexible and then it is easier to find out better

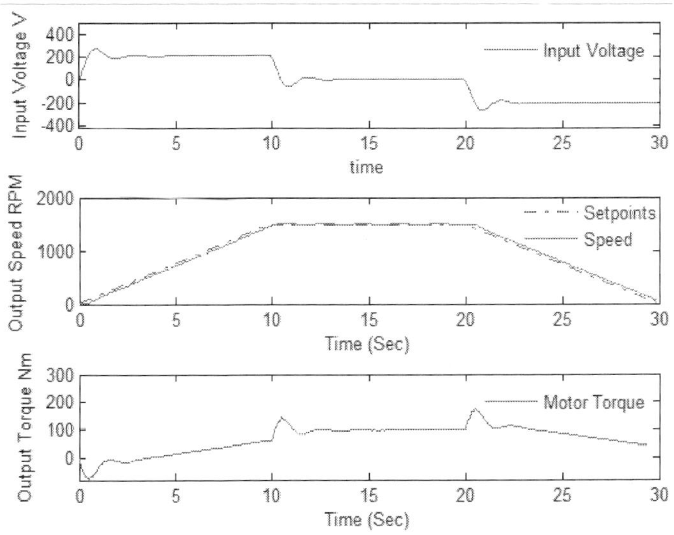

Figure 12. MPC performance with $N_p = 5$ and $Q = 100R$.

solutions. However with the longer prediction, the burden of the computer calculation will exponentially rise up and the time for calculating the optimal actions will depend on the ability of the CPU and the speeds of the communication protocols. In the next simulation, we run the MPC controller with a very shorter horizon length of $N_u = N_y = N_p = 2$; and with medium values of $Q = \begin{bmatrix} 50 & 0 \\ 0 & 50 \end{bmatrix}$ and $R = [1]$. Result of this simulation is shown in Figure 13.

As shown in Figure 13, for very short horizon length prediction of $N_p = 2$. The MPC controller performance becomes worse since the system cannot properly track the output set-points. And then, compared to the next simulation with a longer horizon length prediction, the MPC will achieve a much better performance.

The horizon length prediction is now chosen even longer for $N_u = N_y = N_p = 10$ and for the same values of weighting $Q = \begin{bmatrix} 50 & 0 \\ 0 & 50 \end{bmatrix}$ and $R = [1]$, the performance of this controller is shown in Figure 14.

The MPC scheme with longer prediction length in Figure 14 indicates better performance compared to that in Figure 13. However in this simulation, the much higher input energy is required for the input DC voltages.

A fact is that if we impose an input constraint for voltage of $|V_2| \leq 300V$ and because that the DC motor basically runs in only positive voltage or $0 \leq V_2 \leq 300V$. The braking and the recharged system will be started if we want to slow down the vehicle speed. In the next example, we will

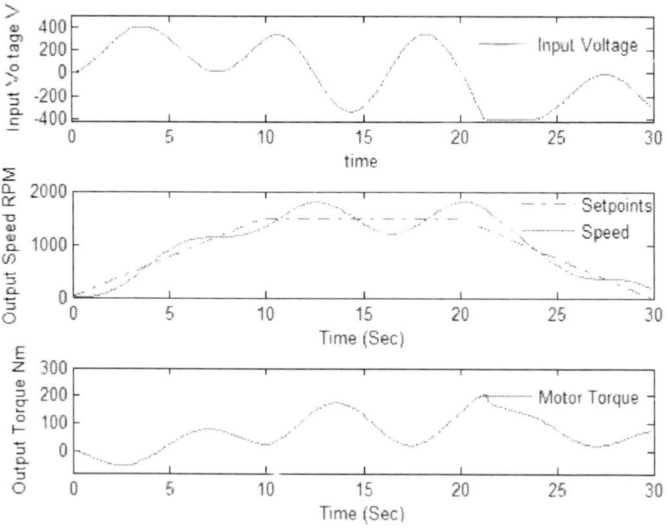

Figure 13. MPC performance with horizon length $N_p = 2$ and $Q = 50R$.

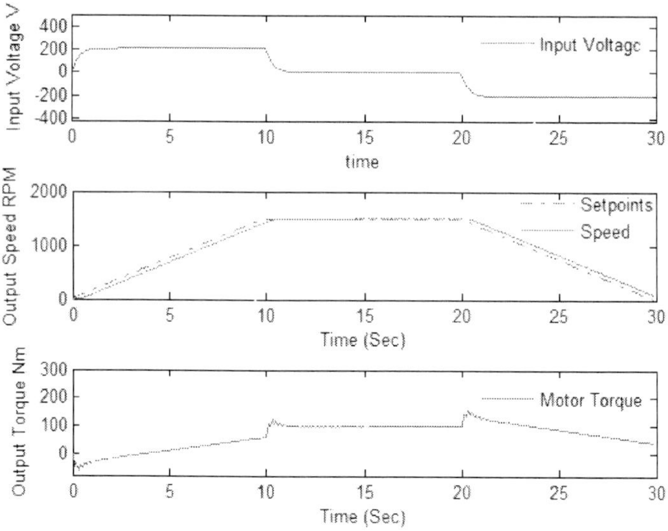

Figure 14. MPC performance with horizon length $N_p = 10$ and $Q = 50R$.

set the input voltage for the DC motor on only positive values of $0 \leq V_2 \leq 300V$. This simulation allows for the regenerative braking option when we start braking to slow down the speed. The performance of this simulation is shown in Figure 15 for the horizon length of $N_u = N_y = N_p = 5$ and high weighting values of $Q = \begin{bmatrix} 1000 & 0 \\ 0 & 1000 \end{bmatrix}$ and small weighting value of $R = [1]$.

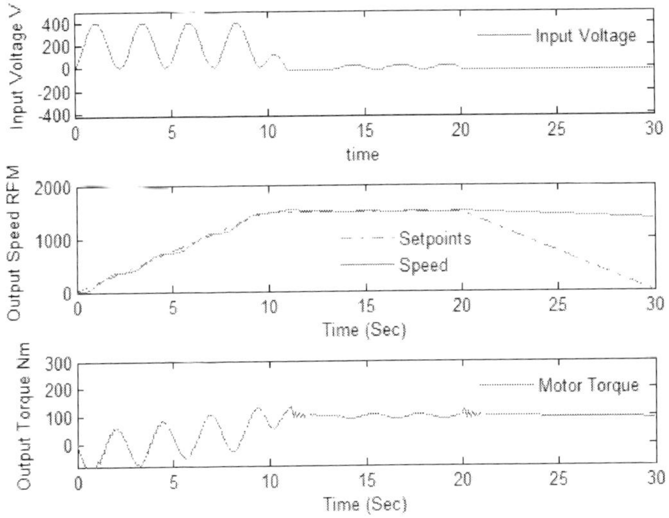

Figure 15. MPC performance with $0 \leq V_2 \leq 300V$, $N_p = 5$ and $Q = 1000R$.

Simulations for the MPC performance with the motor EM2 have been conducted with the change of parameters in the weighting matrices, horizon prediction length, and with the input voltage constraints. For the next part, we will investigate the MPC controller for the ICE and the EM1 to track the desired speed set-points applied for the vehicle high speeds (above 50 km/h) and investigate how to synchronize the velocities of these two parts with the friction clutch engagement.

4.2. MPC performance for ICE and EM1

If the speed of the vehicle exceeds 50 km/h, EM1 will start and activate the ICE to drive the vehicle. Depending on the needed requirements of the speeds and the torques, this hybrid vehicle can configure to operate only the ICE or all ICE, and/or EM1 and/or EM2. For the starting time, the friction clutch is still open and the dynamic formula for the vehicle in the ICE and EM1 is indicated as in equation (27).

The example parameters used for the EM1 are: motor constants, $k_{E2}=k_{T2}=15$; motor inertia, $J_1=1$; motor damping, $k_{\beta 1}=0.5$; armature resistance, $R_{l1}=7$; compensation factor, $\varsigma=0.5$. The system is discretized at a time interval of 0.01 sec. The air drag resistance torque at $\omega_3=2000$ rpm is chosen as $M_{v0}=30$ Nm.

The physical input constraints for this vehicle are set for the DC voltage in range of $|V_1| \le 48V$; The input increment constraint is $\Delta u(t) \le 180V/\text{sec}$. The output constraint of torque for the shaft1 is set for $|T_1| \le 628Nm$.

The MPC parameters are chosen for the horizon length prediction of $N_u=N_y=N_p=5$, the weighting values for the matrices of $Q=\begin{bmatrix}10 & 0\\ 0 & 10\end{bmatrix}$ and $R=\begin{bmatrix}1 & 0\\ 0 & 1\end{bmatrix}$. Figure 16 shows the MPC performance at the starting time.

From Figure 16, after a certain delay time of 1 second, the ICE is fully started and after about 2.4 seconds the peed of ICE has reached the set-point of 2000 rpm and operate stably at a power of 6 kW, generating an output torque of 30 Nm (at this moment, the friction clutch is still open).

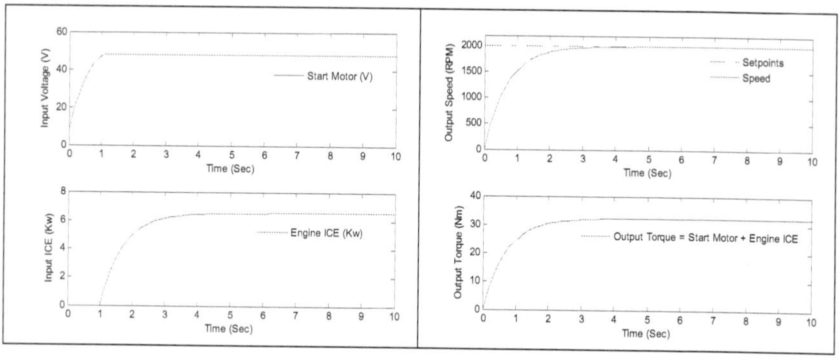

Figure 16. MPC performance at the starting time.

Next, we run the hybrid vehicle with this EM1 and ICE to track the speed set-points and test for the clutch engagement. It is assumed that if the motor EM2 operating at more than 1500 rpm, the motor EM1 will start and activate the ICE to engage to the system. Regarding the improvement of the driving comfort and the reduction of jerk, the clutch engagement will be taken place for only when $\omega_1 \ge \omega_2$ or $\omega_1=1.05.\omega_2$ or the EM1 and ICE must track the EM2 at 5% positive offset. The MPC objective function in (4.3) is now changed from setpoints $r(t)$ to track $\omega_2(t)$ with 5% positive offset. Results of the simulation are shown in Figure 17. The system reaches the setpoint and ready for the clutch engagement after 2.5 seconds.

In the simulation in Figure 17, we operate both motor EM1 and the ICE to track motor EM2 and we can observe that after around 2 sec, the speed of the left hand side clutch disk has exceeded more than 5% of the speed on the right hand side disk and ready for the clutch

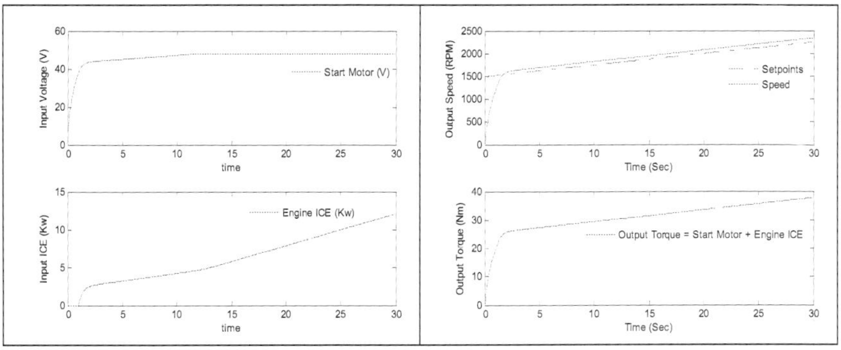

Figure 17. MPC controller for tracking setpoint with both EM1 and ICE.

engaged. However, the motor EM1 functions only for as the ICE starter and once the ICE is fully run, the EM1 can be turned on to as an electrical generator to charge the batteries. In the next example, we will turn off the motor EM1 and use only the ICE to track the speed of the motor EM2. The MPC objective function becomes now similar to the example that we have simulated in Figure 17. Results of the performance are illustrated in Figure 18. The vehicle will reach the speed set-point and ready for the clutch engaged after around 4.4 sec.

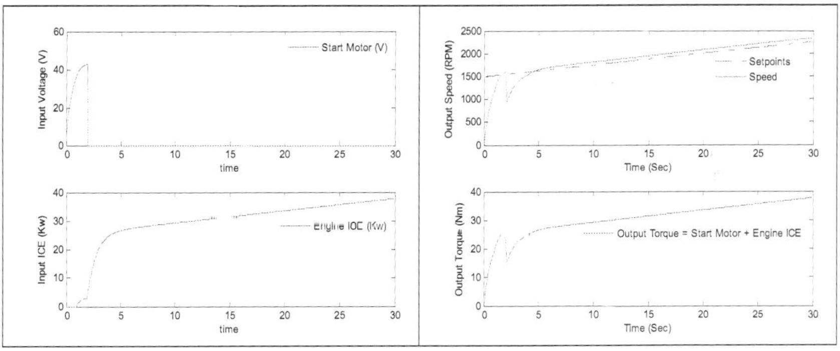

Figure 18. MPC controller for tracking setpoint with only ICE.

How to control the ω_1 rapidly tracking $\omega_2 + 5\%$ by MPC controller is still a big challenge. Next, we test a new MPC controller using the soften output constraints when we consider the output tracking setpoints as in (4.5) with some additional penalty terms added into the MPC objective function, $J_{Soften} = J_{Hard} + \Lambda\varepsilon_i^2(t)$. Results of the simulation are shown in Figure 19.

In Figure 19, additional penalty terms ε_i and a new weighting matrix Λ can be regulated independently together with values of matrices Q and R to obtain some good soften constraints

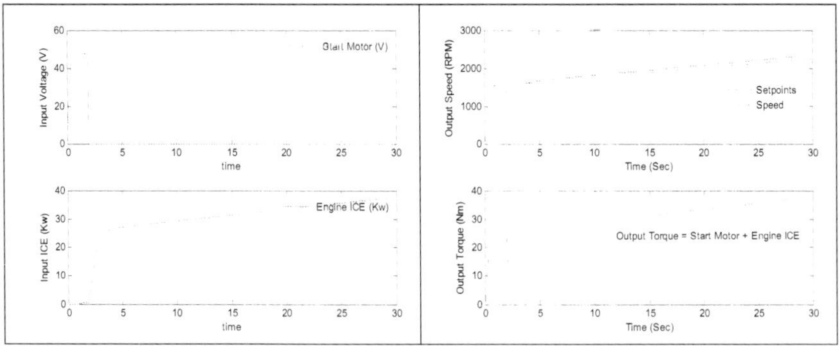

Figure 19. Soften output constraints for tracking set-point with only ICE.

controller performance. The MPC controller becomes looser, more flexible with more regulated parameters. The new system reaches the set-point faster and ready for the clutch engagement after only about 3.4 sec.

5. Conclusions

As a main type of hybrid vehicle, HEVs have achieved better fuel economy and performances. Modern HEV can also improve the efficiency by using the energy from braking and bring other potential environmental benefits. The electric vehicle charging stations can use the low cost and green energy sources from GRID, wind and solar. Due to rapid development in battery technology, the normal electric recharging time has been reducing significantly from 8 h to less than 2 h. The fast recharging time for a modern electric vehicle is now reduced to less than 10 min. Hybrid electric vehicle technology has been applied now not only for the passenger cars but also for all heavy buses and trucks.

The new modeling and control strategy for HEV using MPC has been developed. Reason for using this new control strategy is that, firstly, MPC can solve the optimization problems online with both linear and nonlinear systems, and secondly, MPC can deal with the constraints in the open-loop optimal control problems. MPC can find real-time solution for general constrained nonlinear models over a finite predictive horizon length. Therefore, the performances of the hybrid vehicle can be significantly improved.

The new HEV dynamic modeling equations allow having better studied views for the acceleration of HEVs and the jerk reduction during the transitional engagement period. Examples show that MPC controllers can control the speeds very well to track to any desired speeds. Examples also indicate that the MPC controller can be able to achieve fast and smooth engagement of clutch. The MPC performance can also be considerably improved when we select some appropriate prediction lengths and the values of the weighting values. MPC controller can provide online the optimal control actions subject to the input voltages and output torque constraints. The MPC modified schemes can improve the system performance robustness if some output torque constraints can be softened or turned into the constrained regions.

Acknowledgements

The authors would like to thank the Tallinn University of Technology (TTU) for supporting the preparation of this research article.

Author details

Vu Trieu Minh*, Alina Sivitski, Mart Tamre and Igor Penkov

*Address all correspondence to: trieu.vu@ttu.ee

Department of Mechatronics, Faculty of Mechanical Engineering, Tallinn University of Technology, Tallinn, Estonia

References

[1] Langari R.; Won J.S., A driving situation awareness-based energy management strategy for parallel hybrid vehicles, Proc. Future Transportation Technol Conf, California, 2003; 1; 95-102.

[2] Graaf van der R; Daniel B; Spijker E., Integration of drive system, subsystems and auxiliary systems of a flywheel hybrid driveline with respect to design aspects and fuel economy, Hybridantriebe: Tagung Garching, VDI-Verlag, Düsseldorf, 1999, 1, 459-67.

[3] Lukic S.M.; Emadi A., Effects of drivetrain hybridization on fuel economy and dynamic performance of parallel hybrid electric vehicles, IEEE Transactions on Vehicular Technology, 2004, 53 (2), 385-89.

[4] Jalil N.; Kheir N.A.; Salman, M., A rule-based energy management strategy for a series hybrid vehicle, Proc. American Control Conference, Albuquerque, 1997, 1, 689-93.

[5] Schouten, N. J.; Salman, M. A.; Kheir, N. A., Fuzzy logic control for parallel hybrid vehicles, IEEE Transactions on Control Systems Technology, 2002, 10 (3), 460-68.

[6] Kim, T.S; Manzie, C.; Sharma, R., Model predictive control of velocity and torque split in a parallel hybrid vehicle, Proc. IEEE International Conference on Systems, San Antonio, 2009, 13 (2), 2014 - 19.

[7] Paganelli, G.; Delprat, S.; Guerra, T.M.; Rimaux, J.; Santin, J.J., Equivalent consumption minimization strategy for parallel hybrid powertrains, Proc. IEEE Vehicular Technology Conference, Columbus, 2002, 4, 2076-81.

[8] Sciarretta, A.; Back, M.; Guzzella, L., Optimal control of parallel hybrid electric vehicles, IEEE Transactions on Control Systems Technology, 2004, 12 (3), 352–63.

[9] Qin, S.J.; Badgwell, T.J., An overview of industrial model predictive control technology, AIChE Symposium Series, 1997, 316, 232 56.

[10] Findeisen, R.; Diehl, Z.; Nagy, F.; Allgower, H.; Schloder, J., Computational feasibility and performance of nonlinear model predictive control schemes, Proc. International Conference on Chemical Process Control, Arizona, 2001, 1, 454-60.

[11] Rawlings, J., Tutorial overview of model predictive control, *IEEE Contr. Syst. Mag.*, 2000, 20 (3), 38-62.

[12] Morari, M.;Lee, J., Model predictive control: Past, present and future, Comput. Chem. Eng., 1999, 23 (4), 667-82.

[13] Mayne, D.; Michalska, H., Robust receding horizon control of constrained nonlinear systems, Proc. IEEE *Conference on Decision and Control*, 1993, 38, 1623-33.

[14] Chen, H.; F. Allgower, Nonlinear Model Predictive Control Schemes with Guaranteed Stability, Kluwer Academic Publishers, New York, 1998, 465-94.

[15] Minh, V.T.; Afzulpurkar, N., A comparative study on computational schemes for nonlinear model predictive control, Asian Journal of Control, 2006, 8 (4), 324-31.

[16] Minh, V.T.; Afzulpurkar, N., Robust model predictive control for input saturated and soften state constraints, Asian Journal of Control, 2005, 7 (3), 323-29.

[17] Minh, V.T.; Afzulpurkar, N., Robustness of model predictive control for ill-conditioned distillation process, International Journal of Developments in Chemical Engineering and Mineral Processing, 2005, 13 (3/4), 331-16.

[18] Karden, E.; Ploumen, S.; Fricke, B.; Miller, T.; Snyder, K., Energy storage devices for future hybrid electric vehicles, Journal of Power Sources, 2007, 168, 2-11.

[19] Fredriksson, J.; Egbert, B., Nonlinear control applied to gearshifting in automated manual transmissions, Proc. IEEE Conference on Decision and Control, Sydney, 2000, 1, 444-49.

[20] Powell, B. K.; Bailey, K. E.; Cikanek, S. R., Dynamic modelling and control of hybrid electric vehicle powertrain systems, IEEE *Control Systems Magazine*, 1998, 18 (5), 17-33.

[21] Minh, V.T.; Afzulpurkar, N.; Muhamad, W., Fault detection and control of process systems, Mathematical Problems in Engineering, 2007, Article ID 80321, 20 pages.

[22] Guzzella, L.; Sciarretta, A., Vehicle Propulsion Systems, Introduction to Modeling and Optimization; Springer, Zurich, 2007.

[23] Chan, C. C., The state of the art of electric, hybrid, and fuel cell vehicles, Proceedings of the IEEE, 2007, 95 (4), 704-18.

[24] Miller, J. M., Propulsion Systems for Hybrid Vehicles. IET. 1st edn., London, 2008.

[25] Sciaretta, A.; Guzzella, L., Control of hybrid electric vehicles, IEEE Control Systems Magazine, 2007, 27 (2), 60-70.

[26] Ripaccioli, G.; Bernardini, D.; Di Cairano, S.; Bemporad, A.; Kolmanovsky, V., A stochastic model predictive control approach for series hybrid electric vehicle power management, Proc. American Control Conference, Baltimore, 2010, 1, 5844–49.

[27] Desai, C.; Williamson, S., Comparative study of hybrid electric vehicle control strategies for improved drivetrain efficiency analysis, Electrical Pwoer & Energy Conference, Montreal, 2009, 1, 1-6.

Chapter 3

Fault Diagnosis of Switched Reluctance Motors in Electrified Vehicle Applications

Yihua Hu, Chun Gan, Wenping Cao and Stephen Finney

Additional information is available at the end of the chapter

http://dx.doi.org/10.5772/61659

Abstract

Electric vehicles (EVs) and hybrid electric vehicles (HEVs) can reduce greenhouse gas emissions while switched reluctance motors (SRMs) are one promising motor technology for EVs. This chapter illustrates the fault diagnosis and fault tolerance operation of SRM-based EVs/HEVs, where high reliability is a vital factor involving human lives. Based on the traditional asymmetric half-bridge topology for SRM drives, the characteristics of switching devices upon open-circuit and short-circuit are analyzed, and the corresponding fault diagnosis methods are developed. In order to achieve fault tolerance operation, the central point of SRM stator winding is tapped to form a modular half-bridge configuration to provide fault diagnosis and fault tolerance functions. The fault diagnosis functions are set idle in normal conditions. Simulation results in Matlab/Simulink and experimental results on a 150-W four-phase 8/6 SRM are used to validate the fault identification, and on a 750-W, three-phase 12/8 SRM are used to validate the fault tolerance operation of the proposed strategy, which may have significant implications for EV/HEV applications.

Keywords: Electrified vehicles, fault tolerance, motor drives, switched reluctance motors

1. Introduction

Currently, electric vehicles (EVs) and hybrid electric vehicles (HEVs) represent the future of green transportation and thus are under extensive development across the world [1-3]. In terms of motor drive topology, high-performance permanent magnet (PM) motors are advantageous due to their high efficiency and high torque density [4, 5], but unsustainable for mass production market such as EVs/HEVs because of the scarcity of rare-earth materials they rely on. Considering this reason, more efforts have been devoted to the development of rare-earth-free motor or rare-earth-less motor for future EVs and HEVs [6, 7]. In contrast, switched reluctance motors (SRMs) are becoming a mature technology and are considered to have commercial potentials in widespread applications due to their rare-earth-free feature and wide-range torque-speed characteristics. SRMs have the advantages of robust mechanical structure, low cost, high efficiency, and a wide speed range [8-10]. Hence, the SRM drives have been considered as an attractive solution for the drivetrains of EVs and HEVs [11, 12]. However, to promote the application of SRMs in EVs and HEVs, the following two issues should be properly addressed: 1) the reduction of the weight, complexity, and cost of the drives; and 2) the improvement of the performance and reliability.

In order to satisfy the mentioned requirements, this chapter focuses on fault diagnosis and fault tolerance operation; a new modular fault tolerant topology is proposed on the basis of the traditional SRM driving topology; and the corresponding fault diagnosis and fault tolerance schemes are proposed by trade-off hardware and software.

1.1. Principle of SRM

The structure and principle of switched reluctance motors (SRMs) are derived from 1840s, when the researchers realized that motors can operate by taking advantage of magnetic pull in order. However, the development of high-power thyristors makes it possible to further study of SRM until the 1960s, and then the SRM realized its rapid development in the next time.

As a new speed control system, SRM drive system is integrated with power electronic technology, computer control technology, and microelectronics technology. It has the advantages of both ac and dc speed control systems. Due to its simple structure, low cost, fault-tolerant ability, high efficiency, and wide speed range, SRM has been widely applied in new energy electric vehicles, household appliances, and aviation industry and renewable energy.

The SRM driver system is mainly composed of SRM, power converter, controller, position detector, and current detector, as shown in Figure 1. The SRM has double salient poles structure, and the concentrated windings are rolled around its stator, while its rotor is only made up of silicon steel sheets. Figure 2 shows the structure of a conventional four-phase 8/6-pole SRM. The power converter has many kinds of topologies, while the asymmetrical half-bridge converter topology is one of the most commonly used topologies, as shown in Figure 3.

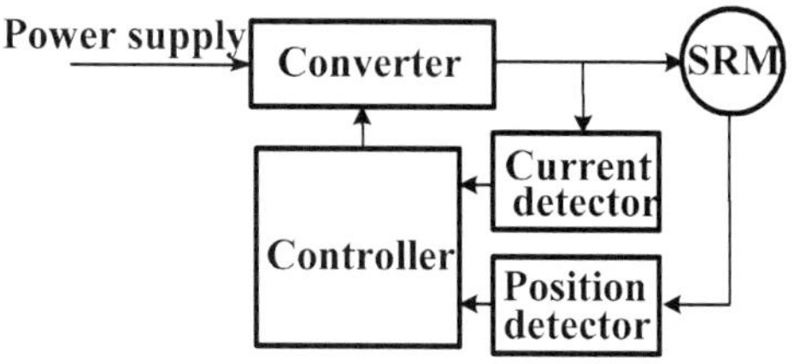

Figure 1. SRM driver system.

http://dx.doi.org/10.5772/61659

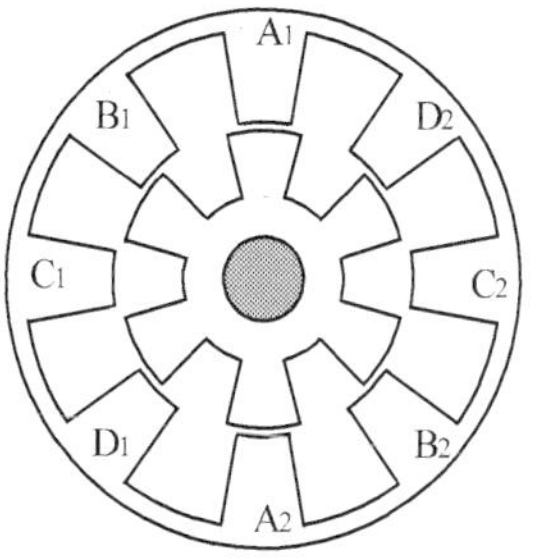

Figure 2. 8/6-pole SRM.

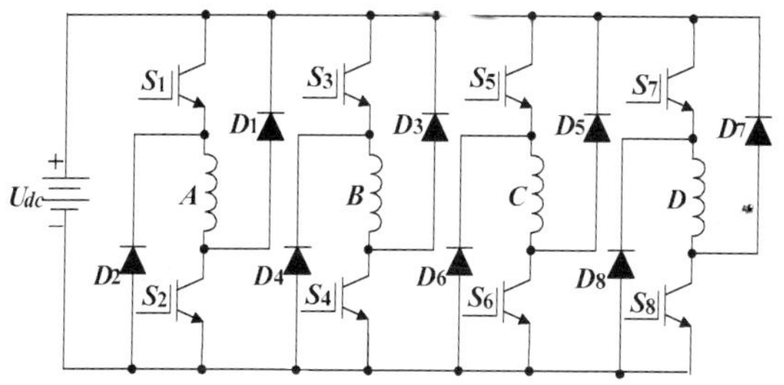

Figure 3. Power converter.

The SRM operation follows the principle that the magnetic flux is always closed along the path of least resistance. When a rotor pole is unaligned with the corresponding stator pole, the magnetic permeance is not at the maximum and the magnetic field can produce a magnetic pull to align with the corresponding poles, as shown in Figure 4. If the conduction sequence of each phase is changed, the motor will rotate inversely. However, the change of phase current direction would not influence the motor rotation.

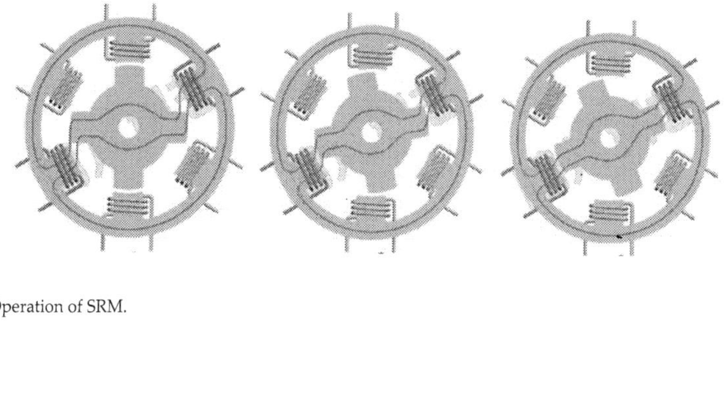

Figure 4. Operation of SRM.

SRM still has some disadvantages due to its special structure and operation mode, such as high torque ripple and noise, and it is difficult to establish the accurate mathematical model. In order to make up for these shortcomings, a lot of technologies have been developed. The optimization of motor structure and the adoption of effective control strategies are used to suppress the torque ripple and noise. And some schemes aim at sensorless control and new power converter topology.

2. Fault diagnosis methods

SRM drives are known to be fault-tolerant by their nature but not completely fault-free. Open-circuit faults are a common fault type of the motor drive, leading to starting difficulty, over-currents, high torque ripples, and reduced load capacity [13, 14]. Due to the harsh condition EVs/HEVs operate at, the switching devices can easily break down. In this section, the fault characteristics of SRM under open-circuit are illustrated; and the fault diagnosis strategy is also presented.

2.1. SRM fault characteristics

Open-circuit faults are common in SRM drives. As shown in Figure 5, there are many locations to initiate an open-circuit fault, such as power converter, motor windings, position signals, and drive circuit. However, the power converter is a crucial component in an SRM drive and is also a vulnerable part in the system [15, 16].

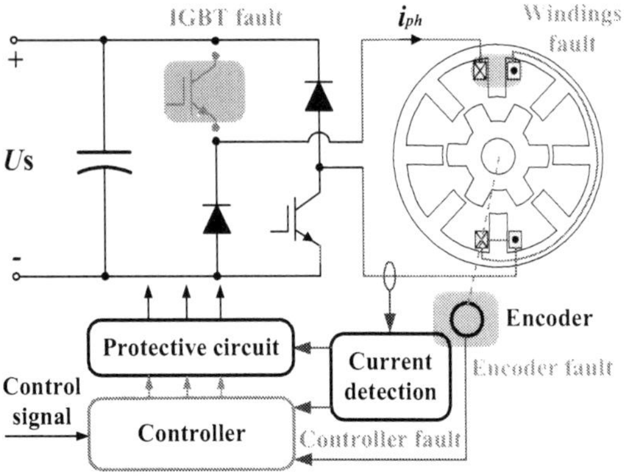

Figure 5. Open-circuit faults for SRM drives.

Each bridge arm is controlled independently by two switches and each phase is independent of each other, providing good fault tolerance. The system will work in the open-circuit state when an open-circuit fault occurs in upper-switch or lower-switch. The average electromagnetic torque under healthy conditions is given by:

$$T_{av} = \frac{mN_r}{2\pi}\int_0^{2\pi/N_r} T_\alpha(\theta,i(\theta))d\theta \tag{1}$$

where m is the phase number, N_r is the number of rotor poles, θ is the rotor angular position, and $T_\alpha(\theta,i(O))$ is the instantaneous torque of one phase. When the open-circuit faults occur at k phases, the faulty phase windings lose excitation but other healthy phases can still operate normally. In this condition, the average torque of the system is given by:

$$T_{fav} = \frac{m-k}{m}T_{av} \tag{2}$$

2.2. Bus current detection

The voltage equation for phase A is given by:

$$U_a = R_a i_a + L_a(\theta)\frac{di_a}{dt} + i_a\omega\frac{dL_a(\theta)}{d\theta} \tag{3}$$

where U_a is the phase voltage, R_a is the phase resistance, L_a is the phase inductance, i_a is the phase current, ω is the motor angular speed, and θ is the rotor position.

The phase current can be written as:

$$i_a(t) = \frac{U_a}{R_a + \omega(dL_a / d\theta)}(1-e^{-t/\tau}) \tag{4}$$

The branch current of phase A is given by:

$$i_{dc\_a} = \begin{cases} i_a & S_1\ close,\ S_2\ close \\ 0 & S_1\ open,\ S_2\ close \\ i_a & S_1\ open,\ S_2\ open \end{cases} \tag{5}$$

For a four-phase motor, the bus current i_{dc} is the sum of four branch currents, given by:

$$i_{dc}(t) = i_{dc\_a}(t) + i_{dc\_b}(t) + i_{dc\_c}(t) + i_{dc\_d}(t) \tag{6}$$

Three bus locations can be defined as follows:

1. When each phase is in the turn-on region $[\theta_{on}\text{-}\theta_{off}]$, the phase voltage is:

$$U_k = \begin{cases} U_s & upper-switch \;\; close, \;\; lower-switch \;\; close \\ 0 & upper-switch \;\; open, \;\; lower-switch \;\; close \end{cases} \tag{7}$$

where U_s is the dc voltage output of the rectifier from the ac power supply. The bus current is in the chopping control state i.e., chopping bus current, denoted by i_{dc2}.

2. When each phase is in the turn-off region $[\theta_{off}\text{-}\theta_{on}+\pi/3]$, the upper-switch and lower-switch are both shut off, and U_k=-U_s. The bus current is in the phase demagnetization state, i.e., demagnetization bus current, denoted by i_{fc}.

3. The bus current in each phase excitation state contains both the chopping bus current and reverse demagnetization bus current, i.e., excitation bus current, denoted by i_{dc1}.

$$i_{dc1} = i_{dc2} - i_{fc} \tag{8}$$

Based on the three bus locations, there are three current sensor placement strategies in the converter, as illustrated in Figure 6. The corresponding detected currents are the chopping bus current in Figure 6(a), the demagnetization bus current in Figure 6(b), and the excitation bus current in Figure 6(c).

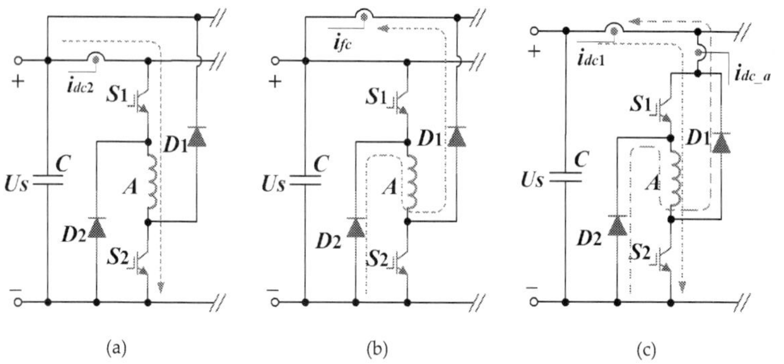

Figure 6. Three bus current detection schemes. (a) Chopping bus. (b) Demagnetization bus. (c) Excitation bus.

2.3. Fault diagnosis based on the FFT algorithm

The FFT algorithm with Blackman window function interpolation is employed to analyze the current spectrums due to smaller side lobe, and better harmonic amplitude accuracy [17].

The fundamental frequencies of the phase current and bus current are given by:

$$f_1 = nN_r / 60 \tag{9}$$

$$f_{bus} = mf_1 = mnN_r / 60 \tag{10}$$

where f_1 is the fundamental frequency of the phase current, f_{bus} is the fundamental frequency of the bus current in normal conditions, n is the motor speed in rpm, and N_r is the number of the rotor poles.

The excitation bus current spectrums before and after the open-circuit faults are shown in Figure 7. There are no harmonic components at f_1 for bus current in normal condition, as shown in Figure 7(a). The dc component A_0 changes little after phase A open, while the phase fundamental frequency component A_{f1}, double-phase fundamental frequency component A_{f2}, and triple-phase fundamental frequency component A_{f3} all increase, especially A_{f1}, as shown in Figure 7(b).

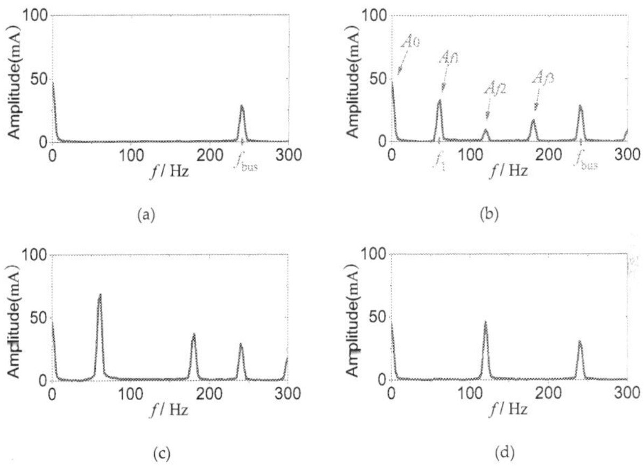

Figure 7. Simulation spectrums of the excitation bus current (a) Normal condition. (b) Phase A open. (c) Phases A and B open. (d) Phases A and C open.

The phase fundamental frequency component A_{f1} and double-phase fundamental frequency component A_{f2} are normalized to the dc component A_0 of the bus current, and the equations are formalized as:

$$A_1^* = A_{f1} / A_0 \tag{11}$$

$$A_2^* = A_{f2} / A_0 \quad (12)$$

where A_1^* and A_2^* are defined as the normalized components of A_1 and A_2.

The simulation results at seven different speeds are shown in Table I. The values of A_1^* and A_2^* are stable at different speeds in the same failure mode. There is no fundamental component in the bus current spectrum under normal conditions; hence, A_1^* is zero when healthy. It increases to 0.7–0.8 when phase A is open, and increases to 1.5–1.7 when phases A and B are open. Similarly, A_2^* rises to about 1.0 when phases A and C are open. Clearly, the normalized components can be used to link with open-circuit faults.

Experimental tests are carried out on a 150-W four-phase 8/6-pole prototype SRM. An asymmetric half-bridge converter is employed in the system, and a DSP TMS320F28335 is used as the main control chip. A fuzzy control algorithm with PWM voltage regulation control is implemented for the closed-loop system. The turn-on angle is set to 0° and the turn-off angle to 28°. An adjustable dc power supply is employed to drive the converter with a 36-V voltage. The IGBT gate signals are controlled to emulate the open-circuit faults.

Speed	Phase A open			Phases A and B open			Phases A and C open		
[r/min]	A_0 [mA]	A_{f1} [mA]	A_1*	A_0 [mA]	A_{f1} [mA]	A_1*	A_0 [mA]	A_{f2} [mA]	A_2*
400	34.63	25.71	0.74	32.52	50.73	1.56	33.66	32.72	0.97
600	47.02	33.12	0.71	45.78	67.30	1.47	44.86	46.25	1.03
800	62.26	47.46	0.76	61.88	100.24	1.62	60.25	58.91	0.98
1000	73.81	56.95	0.77	70.66	122.24	1.63	71.32	68.96	0.97
1200	86.72	65.32	0.75	84.93	139.28	1.64	82.13	78.68	0.96
1400	99.86	74.11	0.74	98.32	155.35	1.58	98.18	96.21	0.98
1500	108.63	83.93	0.77	106.21	176.31	1.66	105.26	102.58	0.97

Table 1. Simulation results of the excitation bus current

Figure8 (a)-(d) shows the chopping bus current i_{dc1}, excitation bus current i_{dc2}, and demagnetization bus current i_{fc} before and after phase A open, phases A and B open, and phases A and C open, respectively, at 600 r/min. The FFT algorithm with Blackman window interpolation is generated on the bus current before and after the fault, ranging from 400 to 1500 r/min.

The phase fundamental frequency component A_{f1} is very small at normal conditions. However, A_{f1} is zero in the simulation due to an ideal condition, while a small harmonic component exists in the phase fundamental frequency because of the electromagnetic interference (EMI) and rotor eccentricity in real conditions, which does not affect the accuracy of the diagnosis. The dc component A_0 does not change and the phase fundamental frequency component A_{f1} increases obviously when faults happen. Hence, A_1^* increases both after phase A open and after phases A and B open.

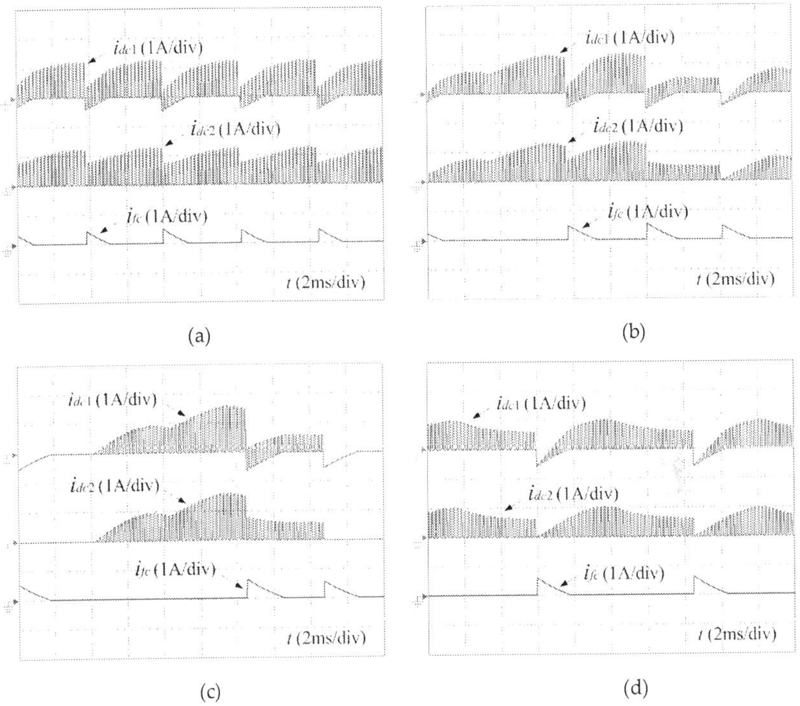

Figure 8. Experimental results of bus currents. (a) Healthy. (b) Phase A open. (c) Phases A and B open. (d) Phases A and C open.

The fault characteristics of the bus current before and after phase A open and phases A and B open are shown in Figure 9. A_1^* of the excitation bus current, chopping bus current, and demagnetization bus current are all below 0.05 at normal conditions, while the values are stable within 1.6–1.7, 0.9–1.0, and 1.4–1.5, respectively, when phases A and B are open-circuited. Clearly, the fault characteristics confirm that the normalized phase fundamental frequency component of the bus current can be used as the open-circuit fault signature.

Under phases A and C open condition, A_1^* cannot be used to diagnose this fault since there is no change in A_{f1} compared to the normal state, as shown in Figure 9(d). However, A_{f2} changes obviously and its normalized component A_2^* can thus be used for this diagnosis. Figure 9(d) shows A_2^* before and after phases A and C open. Curve 1, curve 2, and curve 3 represent A_2^* of the chopping bus current, excitation bus current, and demagnetization bus current, respectively, in normal conditions. Curve 4, curve 5, and curve 6 represent A_2^* of the chopping bus current, excitation bus current, and demagnetization bus current, respectively, when phases A and C are open-circuited. As illustrated in the figure, the stability for curve 6 is poor, and the value declines as the speed increases, while the fault characteristic of curve 5 is more significant and more stable compared to others, which can be used for diagnosis of this fault.

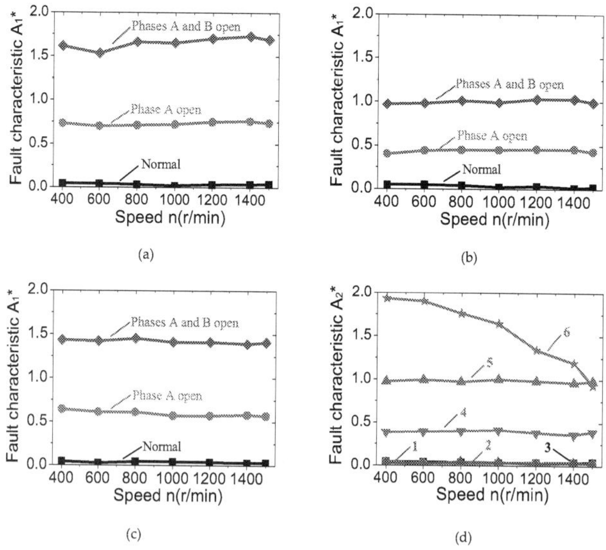

Figure 9. Fault characteristics before and after open-circuit faults. (a) A_1^* of excitation bus. (b) A_1^* of chopping bus. (c) A_1^* of demagnetization bus. (d) A_2^* before and after phases A and C open.

By comparing the three fault characteristics, it becomes clear that the excitation bus current shows a good correlation with the faults. Therefore, it is chosen to extract the fault characteristics for fault diagnosis.

The fault characteristic has good robustness, as shown in Figure 10. A_1^* has good disturbance resistance to the changes in the load and turn-on angle. Therefore, this diagnostic method is suitable for variable load drives and variable angle control systems. It needs to point out that the sampling frequency should be greater than the PWM chopping frequency to ensure the accuracy of the current detection and harmonic analysis.

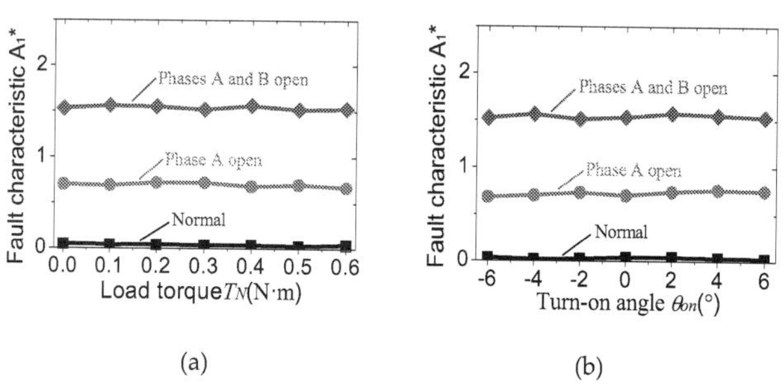

Figure 10. Load torque and turn-on angle in relation to A_1^*. (a) A_1^* with load torque. (b) A_1^* with turn-on angle.

Figure 11 shows the value of A_1^* when the system is subject to a fast transient disturbance. Clearly, A_1^* does not change with the load variations, speed regulation, and angle modulation. Therefore, the proposed method has excellent robustness to the fast transients, which would not generate false alarms.

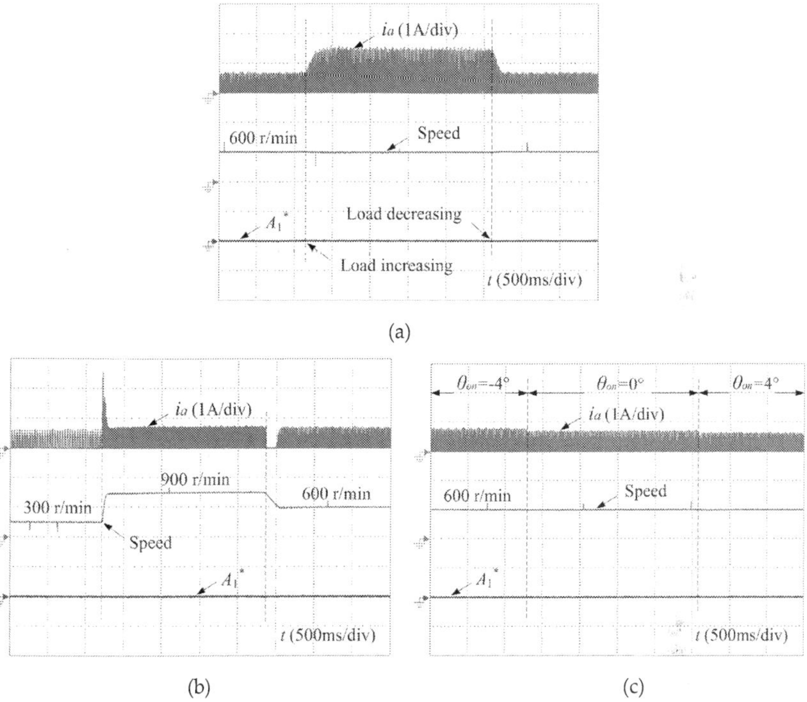

Figure 11. Fast transients in relation to A_1^*. (a) Load variation. (b) Speed regulation. (c) Angle modulation.

3. Fault tolerant topology

Traditionally, the SRM phase windings are composed of an even number of series connected windings, as shown in Figure 12. Thus, central-tapped windings are formed, which can be easily designed in 8/6 or 12/8 SRM. Figure13 shows the traditional 12/8 SRM winding connection, in which L_{a1}, L_{a2}, L_{a3}, and L_{a4} represent for four windings of one SRM phase; the central tapped node A of phase L_a is developed as shown in Figure 13. One phase of SRM drive circuit is composed by traditional asymmetrical half-bridge topology and phase winding; the whole circuit can be divided into two parts: left part and right part; each part has the same components, including diode, switching device, and phase winding, as presented in Figure 14. The two parts have the characteristics of axial symmetry that can be employed in fault tolerance operation. When the central tapped node is connected with positive node of power supply source, the left part of the converter is bypassed, which can block the left part fault. The same method, when the central tapped node A is connected with negative node of power supply source, the right part of the converter is bypassed, which can block the fault from right part.

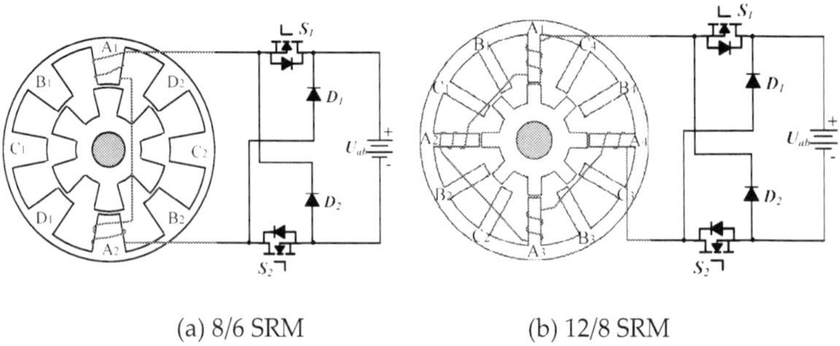

Figure 12. Basic winding structure of SRM.

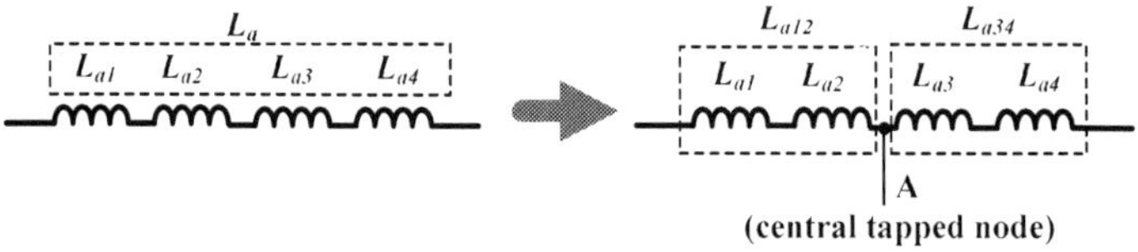

Figure 13. Central-tapped winding of a 12/8 SRM.

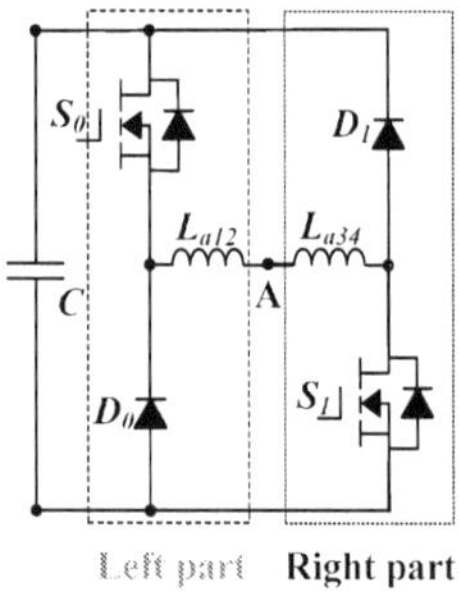

Figure 14. Two parts of phase converter.

http://dx.doi.org/10.5772/61659

On the basis of the central tapped node and axial symmetry characteristics of the traditional drive topology, the proposed fault tolerant topology is presented in Figure15. Figure 15(a) is the main driving topology composed of main topology (traditional asymmetrical half-bridge) as Figure 15(b), and fault tolerance module as Figure 15(c). The fault tolerance module is the traditional three-phase half-bridge modular. The half-bridge central nodes are connected with central tapped node of phase windings, which are A, B, and C, respectively. Three-phase half-bridge is employed to approach fault tolerance operation. The proposed topology has the characteristics of modular structure; on the base of traditional asymmetrical half-bridge topology, only one three-phase bridge modular is needed. The basic structure of SRM is almost not changed. In normal conditions, the proposed topology works as traditional asymmetrical half-bridge topology; the fault tolerance module is in idle condition that makes the proposed converter have the same efficiency as the traditional asymmetrical half-bridge topology. The fault tolerance module works only at fault condition.

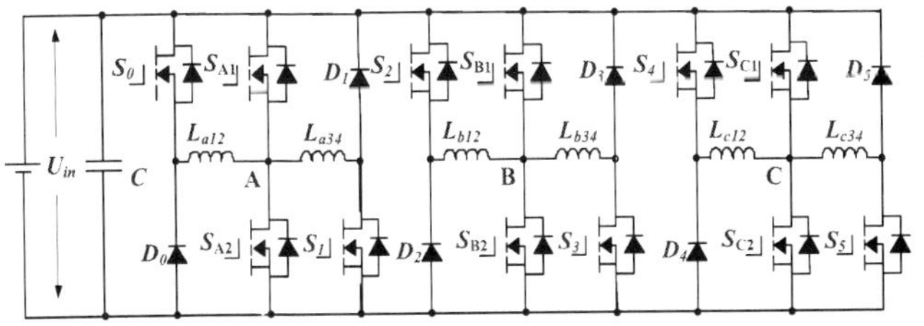

(a) Proposed fault tolerance topology

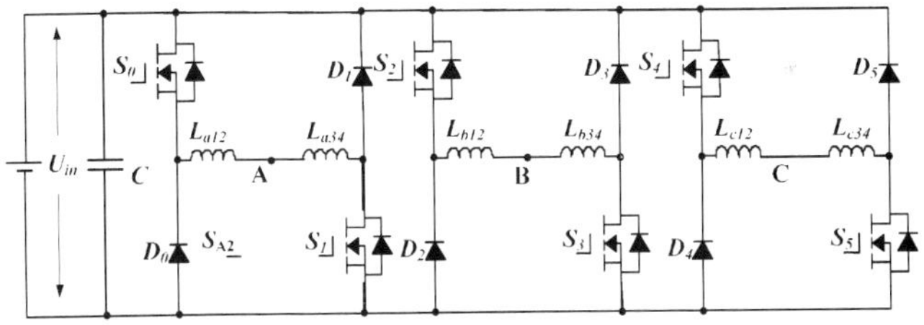

(b) Main topology for SRM

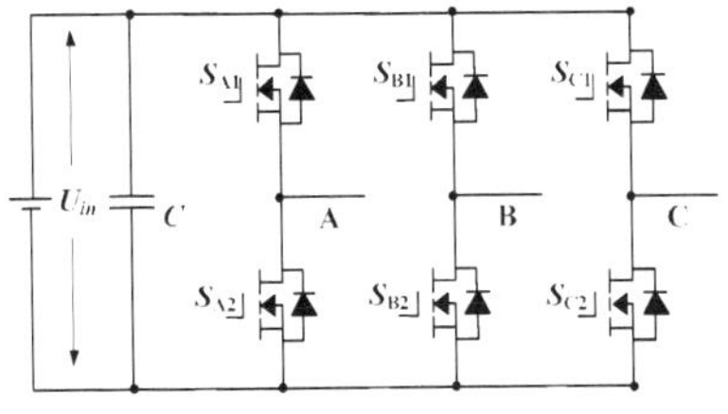

(c) Fault tolerant module

Figure 15. Proposed topology for SRM fault tolerance operation.

3.1. Switching device faults and phase winding open-circuit faults

Switching device faults and phase winding open-circuit faults are common fault phenomena. In the traditional asymmetrical half-bridge converter, there are two switching devices for each phase; and each phase has four windings for a 12/8 SRM. When there is no current in the excitation region in phase L_a, it means that the open-circuit occurs. The diagnosis needs to locate which part is under fault condition by replacing S_0 by S_{A1}, and giving the turn-off single to S_0. In the right part of the converter, S_{A1}, S_{A2}, D_1, S_1, and L_{a34} compose a new asymmetrical half-bridge. In the right part asymmetrical half-bridge, if the faulty phase can work, it proves that the left part of converter is under fault condition. By the same method, replacing S_1 by S_{A2}, and giving the turn-off single to S_1; in left part of converter, S_0, S_{A1}, S_{A2}, D_0, and L_{a12} compose a new asymmetrical half-bridge. In the left part asymmetrical half-bridge, if the faulty phase can work, it proves that the right part of the converter is under fault condition. The diagnosis flowchart of the open-circuit fault is shown in Figure 16.

3.2. Fault tolerance operation under open-circuit fault conditions

When a faulty part is identified in the faulty phase, the fault tolerance module and main topology combines new topology for faulty phase converter. If the left part of one phase converter is in fault condition, in the fault tolerant topology, the left part of one phase converter, including the switching device S_0, diode D_0 and phase winding L_{a12}, is shorted by half-bridge S_{A1} and S_{A2} to block faulty part. Figure 17(a) is the typical example of S_0 under open-circuit condition. When open-circuit fault occurs, the half-bridge S_{A1} and S_{A2} is activated to combine with right part converter to form a new fault tolerant topology. In the new-formed fault tolerance operation topology, when S_{A1} and S_1 conduct, the excitation circuit is shown in Figure 17(b). Figure 17(c) presents the energy-recycling mode, in which the winding voltage is $-U_{in}$ to speed up winding demagnetization. Figure 17(d) shows the freewheeling conduction mode, in which the winding voltage is 0. The working modes of the fault tolerance converter are the same as a traditional converter, except only a half phase-winding is operating.

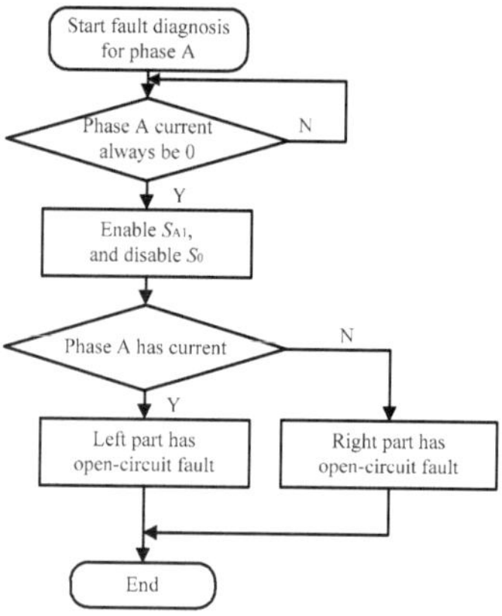

Figure 16. Flowchart for the diagnosis of the open circuit fault.

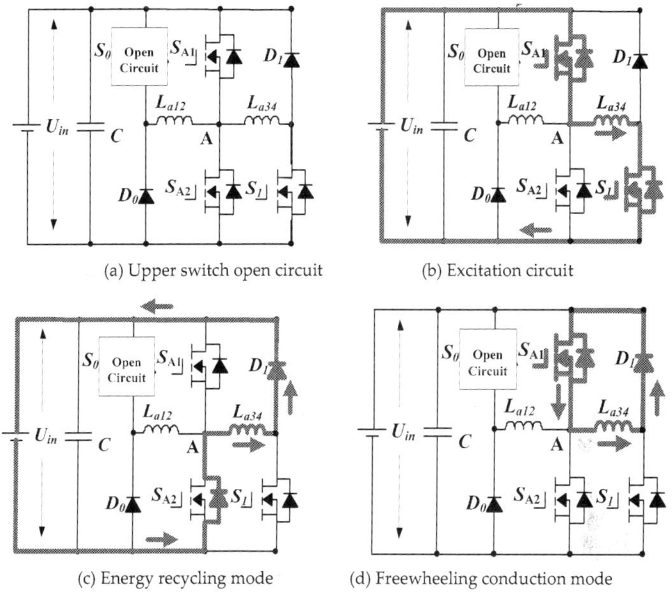

Figure 17. Fault tolerance operation topology under open-circuit condition.

The same fault tolerance operation can be achieved, when D_0 or L_{a12} is faulty. When the right part of one phase converter is under open-circuit fault condition, the left part of one phase converter and fault tolerance module can combined to form a new fault tolerant topology, using the same method as Figure 17.

3.3. Switching device and phase winding short-circuit fault diagnosis

When S_0 is short-circuited, a freewheeling loop is formed, as shown in Figure 18(a). When the short-circuit fault of switching device S_1 occurs, the only freewheeling mode is illustrated in Figure 18(b); both in Figure 18(a) and 18(b) fault condition, the corresponding phase current is always over zero, which can be employed in short-circuit fault diagnosis.

When a short-circuit fault is detected, the next step is to locate which switching devices is under fault condition. For example, if S_0 has a short-circuit fault, S_{A1} is enabled by giving drive signals to replace S_0. The right part of converter and half-bridge compose a new asymmetrical half-bridge. In the right-part asymmetrical half-bridge, if the freewheeling current can decrease to zero, the faulty part can be located in S_0; the right-part converter and half-bridge can form a new converter to achieve fault tolerance operation that is the same as open-circuit. The diagnosis flowchart of the short-circuit fault is shown in Figure 18(c).

The left part of the converter and half-bridge compose a new asymmetrical half-bridge. In the left-part asymmetrical half-bridge, if the freewheeling current can decrease to zero, the faulty part can be located in S_1; the left-part converter and half-bridge can form a new converter to achieve fault tolerance operation that is the same as open-circuit.

For instance, when the switching device S_0 is short-circuited, the half-bridge arm and right-part converter form new topology. In order to block the faulty part, switching devices S_1 is employed as chopping devices. In excitation and freewheeling state, due to S_{A1} conducting, both sides of phase winding L_{a12} share the same electric potential, which prevents the current forming in L_{a12}; in energy recycle mode, there is no current loop for L_{a12}. Therefore, in the three basic working states, there is no current in phase winding L_{a12}, as presented in Figure 19(a)-(c),which proves that left-part converter is blocked. Similarly, when switching device S_1 is in short-circuit condition, in order to block the right-part converter, S_0 is employed as chopping switch.

Inner turn short-circuit faults are also the faults to cause the decreasing of phase inductance; but the faulty phase still can operate. The proposed fault tolerance strategy also can bypass the short-circuited part to stop it from propagating.

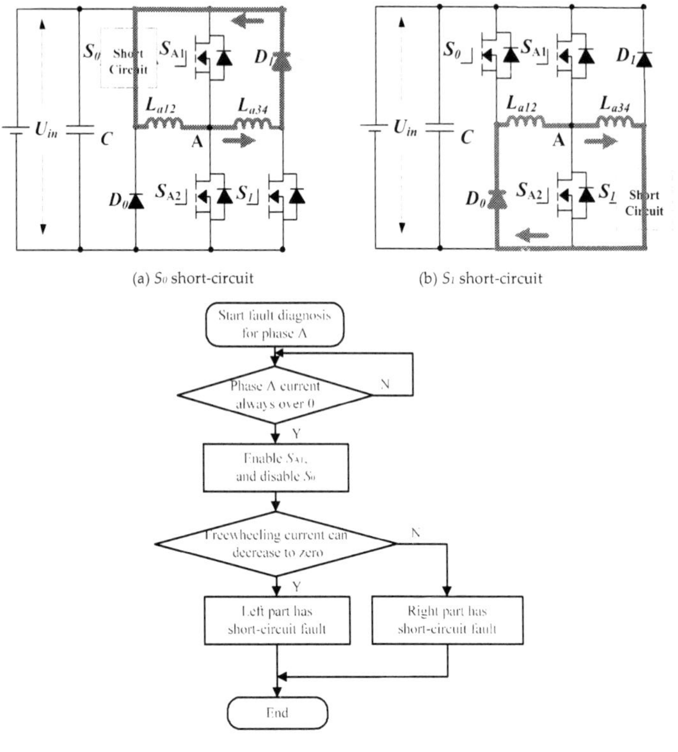

Figure 18. Diagnosis of switching device short-circuits.

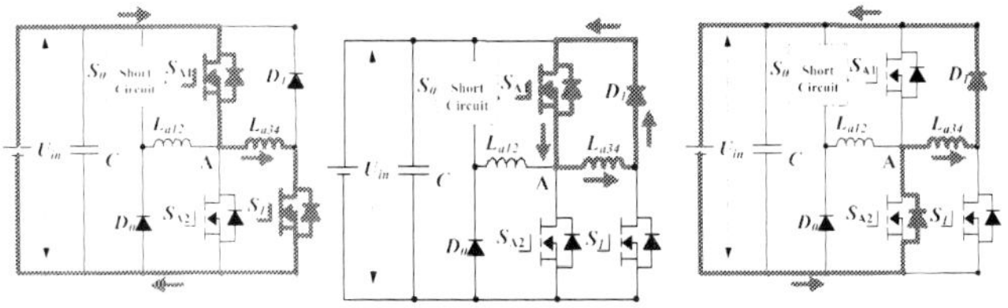

(a) Excitation circuit (b) Freewheeling conduction mode (c) Energy recycling mode

Figure 19. Fault tolerance operation under S_0 short-circuit condition.

3.4. Fault tolerance operation control strategy

After locating a fault part, the corresponding fault tolerance control strategy is needed to deal with the faulty condition.

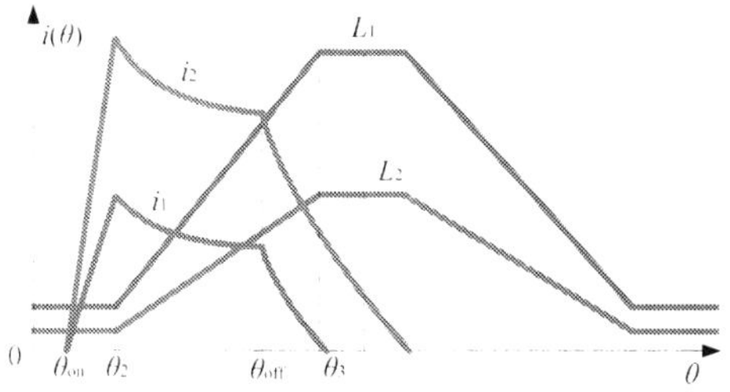

(a) Fault tolerance operation with a half phase winding

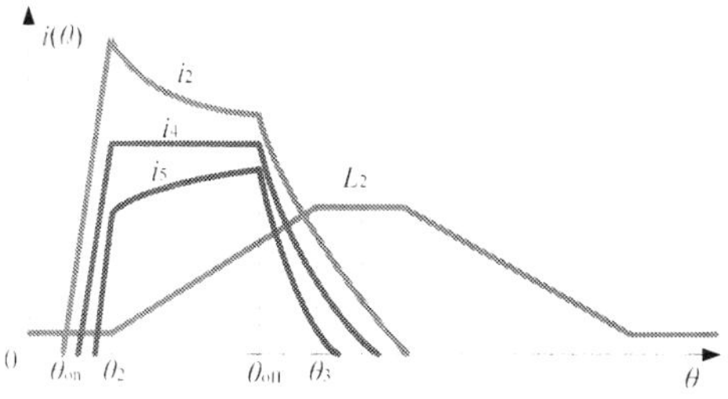

(b) Turn-on angle lagging behind

Figure 20. Relationship between the phase current and phase inductance.

Figure 20 shows the relationship between phase current and phase inductance. As illustrated in the figure, θ_{on} and θ_{off} are the turn-on and turn-off angles, respectively, i_1 and L_1 are the phase current and phase inductance under the normal conditions, i_2 and L_2 are under the fault tolerance conditions, and i_4 and i_5 are the phase currents when the turn-on angle is set lagging behind. Figure 20(a) shows the phase current and phase inductance in the fault tolerance operation with a half phase winding, compared to a normal operation. Figure 20(b) shows that the phase current operates in fault tolerance conditions when the turn-on angle is set lagging behind.

The phase inductance slope factor in the inductance ascending region is expressed as:

$$K_L = \frac{L_{\max} - L_{\min}}{\theta_3 - \theta_2} \tag{13}$$

where $L_{\min}$ and $L_{\max}$ are the minimum and maximum of the phase inductance, and θ_2 and θ_3 are the corresponding rotor position.

In the region of $\theta_{on} \leq \theta < \theta_2$, the phase current is expressed as:

$$i(\theta) = \frac{U_{in}}{\omega_r} \frac{\theta - \theta_{on}}{L_{\min}} \tag{14}$$

where U_{in} is the bus voltage, and ω_r is the angular velocity. In this region, the phase current goes up quickly, following the current slope factor K_i, given by:

$$K_i = \frac{di}{d\theta} = \frac{U_{in}}{\omega_r L_{\min}} > 0 \tag{15}$$

In the region of $\theta_2 \leq \theta < \theta_{off}$, the phase current is expressed as:

$$i(\theta) = \frac{U_{in}}{\omega_r} \frac{\theta - \theta_{on}}{L_{\min} + K_L(\theta - \theta_2)} \tag{16}$$

The peak value of the phase current is at the position $\theta = \theta_2$, which is given by:

$$i_{max} = \frac{U_{in}}{\omega_r} \frac{\theta_2 - \theta_{on}}{L_{\min}} \tag{17}$$

The average electromagnetic torque of one phase is given by:

$$T_{av} = \frac{N_r}{2\pi} \frac{U_{in}^{\ 2}}{\omega_r^2} \left(\theta_{off} - \theta_2\right) \left(\frac{\theta_2 - \theta_{on}}{L_{\min}} - \frac{1}{2} \frac{\theta_{off} - \theta_2}{L_{\max} - L_{\min}} \right) \tag{18}$$

where N_r is the rotor poles.

If the motor system has an open-circuit or short-circuit fault in the converter, the proposed converter will operate with a half part of the fault phase winding, then

$$\begin{cases} L_{\max}' = \frac{1}{2} L_{\max} \\ L_{\min}' = \frac{1}{2} L_{\min} \end{cases} \quad (19)$$

where $L_{\min}'$ and $L_{\max}'$ are the minimum and maximum of the faulty phase inductance.

The phase inductance slope factor in the inductance ascending region in the fault-tolerant operation is as follows:

$$K_L' = \frac{1}{2} \frac{L_{\max}' - L_{\min}'}{\beta_s} = \frac{1}{2} K_L \quad (20)$$

In the region of $\theta_{on} \leq \theta < \theta_2$, the phase current slope in fault-tolerant operation is as follows:

$$K_i' = (\frac{di}{d\theta})' = \frac{U_{in}}{\omega_r \frac{1}{2} L_{\min}} = \frac{2U_{in}}{\omega_r L_{\min}} = 2K_i \quad (21)$$

The peak value of the phase current at the position $\theta=\theta_2$, in fault-tolerant operation, is:

$$i_{max}' = \frac{U_{in}}{\omega_r} \frac{\theta_2 - \theta_{on}}{\frac{1}{2} L_{\min}} = \frac{U_{in}}{\omega_r} \frac{2(\theta_2 - \theta_{on})}{L_{\min}} = 2i_{max} \quad (22)$$

The average electromagnetic torque of the failure phase is given by:

$$T_{av}' = \frac{N_r U_{in}^2}{2\pi\, \omega_r^2} \left(\theta_{off} - \theta_2\right) \left(\frac{\theta_2 - \theta_{on}}{\frac{1}{2} L_{\min}} - \frac{1}{2} \cdot \frac{\theta_{off} - \theta_2}{\frac{1}{2} L_{\max} - \frac{1}{2} L_{\min}} \right) = 2T_{av} \quad (23)$$

According to Eqs. (22) and (23), the peak value of the phase current and the average electromagnetic torque of the failure phase are double of the normal value when working in fault-tolerant operation. However, in a closed-loop system, the total average electromagnetic torque is the same as that in the normal state, due to a constant load.

When an open-circuit fault of the drive happens, conventionally, the system still works in the phase absence operation to ensure the continued working ability in a closed-loop system. However, the currents of other normal phases will be larger than the previous one to compensate the torque output, due to the adjustment of the speed controller. The unbalanced phase current increases the torque ripple and the load capacity also is reduced considerably. When the short-circuit fault of the drive happens, the demagnetization current cannot decrease to zero due to a zero-voltage loop, which causes phase current to become more unbalanced and obviously increases the torque ripple.

Considering the proposed fault tolerance scheme in the CCC system, a half of the failure phase still can be put into use to ensure the torque output. Since the phase current is the control target, it will be regulated to the same reference compared to the normal one, even though a half of the failure phase is removed. In voltage-PWM control strategy, the phase voltage is the control target. The imposed voltage on each phase is the same, regardless of the whole or half of the phase winding works. In order to reduce the unbalanced phase current further in voltage-PWM system, the turn-on angle of the failure phase can be adjusted lagging behind to reduce the increased phase current in the failure winding, as illustrated in Figure 20(b). Hence, the proposed drive topology can be used to compensate the current and torque, and reduce the torque ripple to improve the drive performance in fault conditions.

3.5. Experimental verification

To verify the effectiveness of the proposed scheme experimentally, a test rig for testing a 750-W SRM prototype is set up, as shown in Figure 21(a). Two air switches are adopted to emulate open-circuit and short-circuit faults, as shown in Figure 21(b), where J_1 is used to achieve an open circuit fault and J_2, a short circuit fault. Figure 21(c) shows the fault tolerance control system diagram with the closed-loop speed regulation capability. As illustrated in the figure, a PI controller is used to regulate the motor speed, and the proportional gain and integral gain are 0.05 and 0.5, respectively. The current controller and voltage controller are utilized to generate the drive signals to control the motor drive in different operation modes. The position detector and speed calculator are used to give the instantaneous speed for feedback control. The current sampling and fault diagnosis schemes are employed to control the gate signals for the fault tolerant topology to operate under faulted conditions.

The type of the MOFESTs used is FDA59N30 from Fairchild Inc; and diodes are IDW75E60 from Infineon Technologies. Three current sensors (LA55Ps) are used to measure the phase currents. An incremental encoder with 1000 lines is used to measure the rotor position. A dSPACE 1006 control board is employed to implement the control scheme. A magnetic brake acts as the load with a torque of 1 N·m. The dc-link voltage is fixed to 48V. Two air switches are adopted to generate open-circuit and short-circuit faults. The torque observed in the oscilloscope is obtained online by using the real-time phase currents and rotor position to look up for the torque value in a 3D torque table that includes the T-i-θ characteristics [18, 19]. The torque data in the lookup table are measured by using a rotor-clamping device when supplying different steady currents to the motor windings in a rotor position that changes step-by-step. The output torque in the experimental waveforms is observed through a D/A converter.

The turn-on and turn-off angles are set to 0° and 20°, respectively. In voltage-PWM control system with fault tolerant topology, the turn-on angle is set to 5° to improve the phase current balance for the fault tolerance performance when the short-fault occurs. Figures 22 through 25 present the experimental results at 500 r/min, where i_a, i_b, and i_c are the phase currents for phase A, B, and C, respectively; T^* and T are the given load torque and instantaneous torque, which show a good agreement with the simulation results. Figure 22 presents the typical voltage-PWM control model waveforms of the SRM under normal, open-circuit fault, and short-circuit fault conditions. In a normal condition, three phases have the same current amplitude and shape. In an open-circuit faulty condition, there is no current in the faulty phase. In a short-circuit faulty condition, the faulty phase current cannot decrease to zero. The experiment results have agreed well with the analytical study in Section 3.3. Figure 23 verifies the control strategy under fault condition. By controlling the turn-on angle of phase A, the output torque ripple can be decreased. Figure 24 shows the typical waveforms for the current regulation control model under normal, open fault, and short fault conditions. In an open-circuit faulty condition, there is no current in the faulty phase, while in a short-circuit faulty condition, as theory analysis, the fault phase current cannot decrease to zero. The experiment results also verify the theory analysis in Section 3.2. Under these fault tolerance operation conditions, the faulty phase current and output torque with the proposed method can follow the reference values faithfully, as shown in Figure 25. This is also the case in reducing the torque ripple and the imbalance between phase currents for conventional converters with either open-circuit or short-circuit faults.

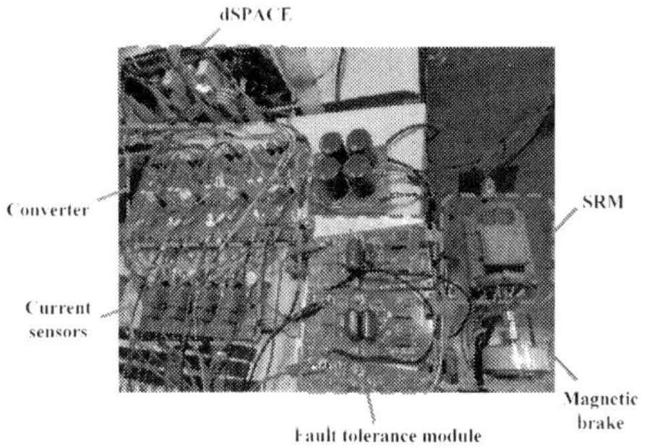

(a) Experimental setup

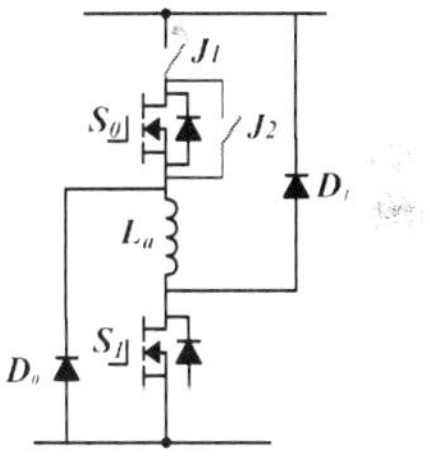

(b) Generation of faults by J_1 and J_2

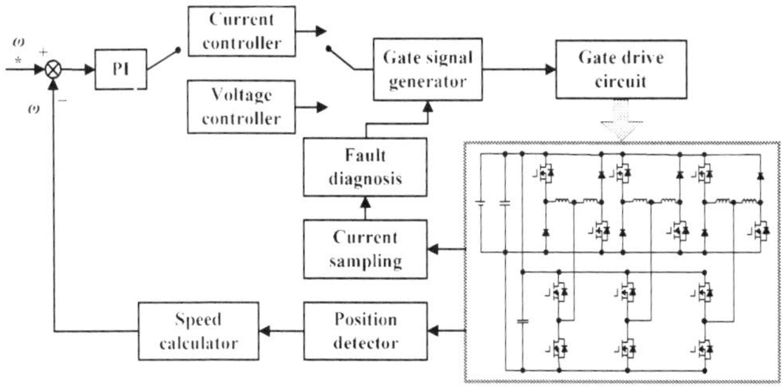

(c) Diagram of the fault tolerance control system

Figure 21. Experimental setup and the control system.

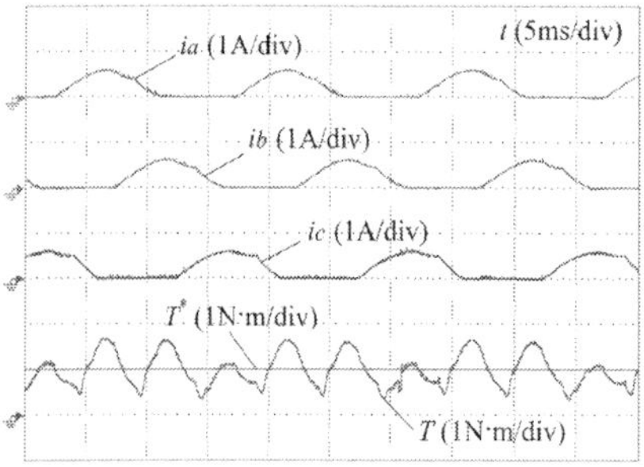

(a) Normal

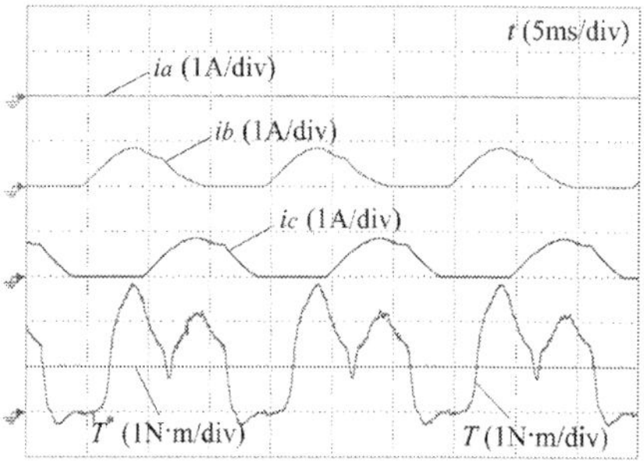

(b) Open fault

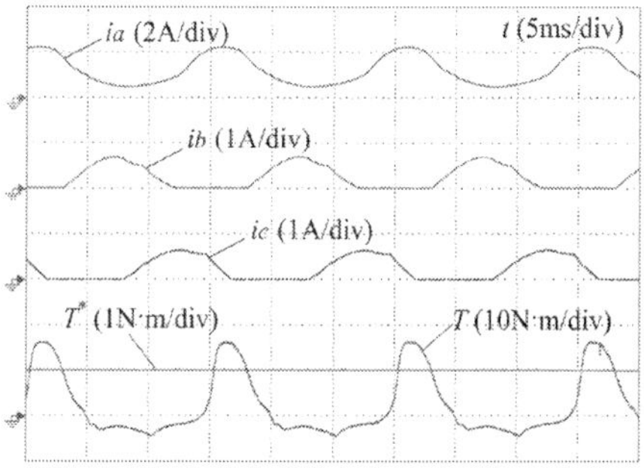

(c) Short fault

Figure 22. Experimental results of voltage-PWM control mode under normal and fault conditions.

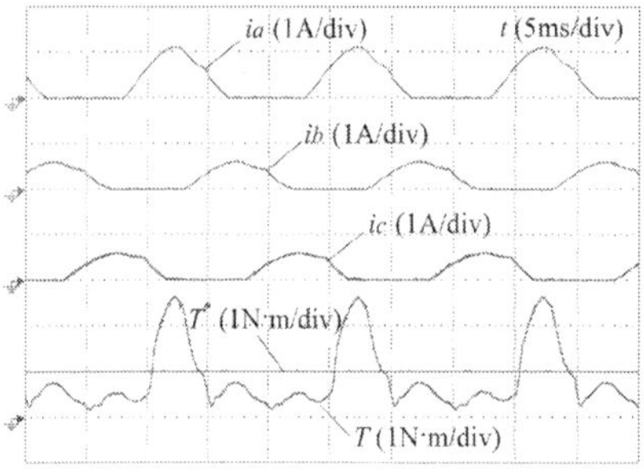

(a) Turn-on angle is 0° for phase A

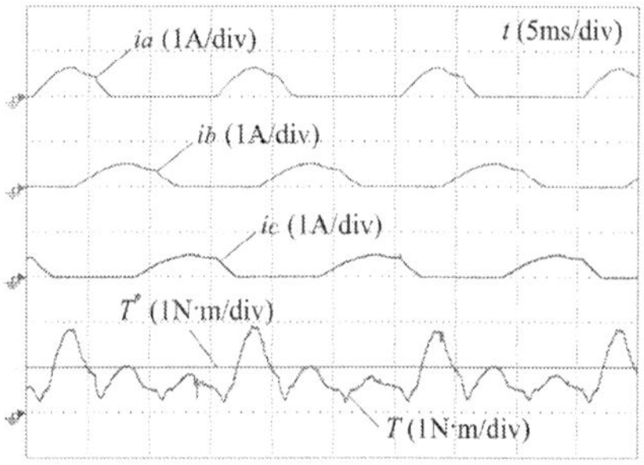

(b) Turn-on angle is 5° for phase A

Figure 23. Experimental results of voltage-PWM control mode with fault tolerant topology under fault conditions.

http://dx.doi.org/10.5772/61659

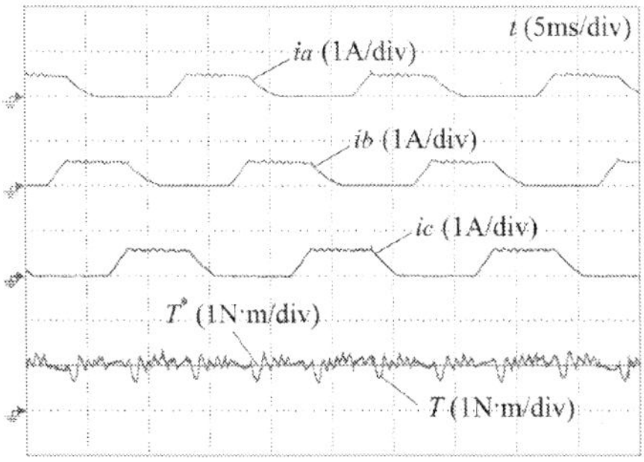

(a) Normal

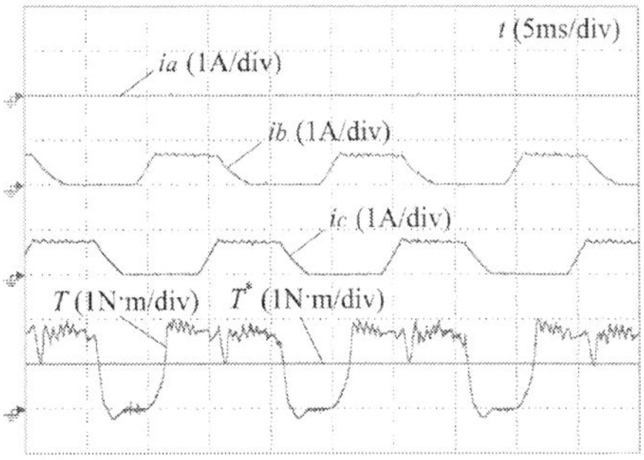

(b) Open fault

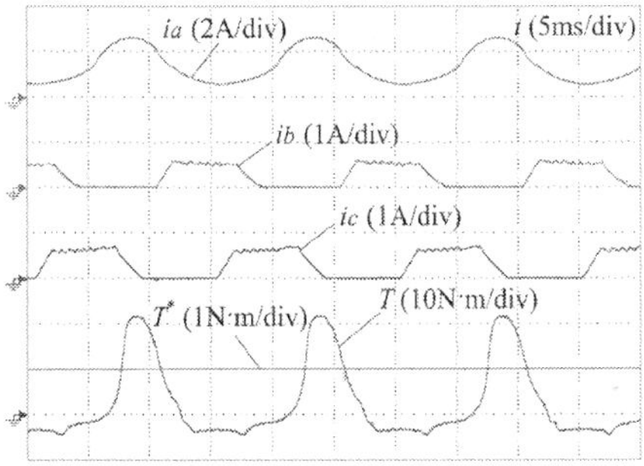

(c) Short fault

Figure 24. Experimental results of current regulation control mode without fault tolerant topology under fault conditions.

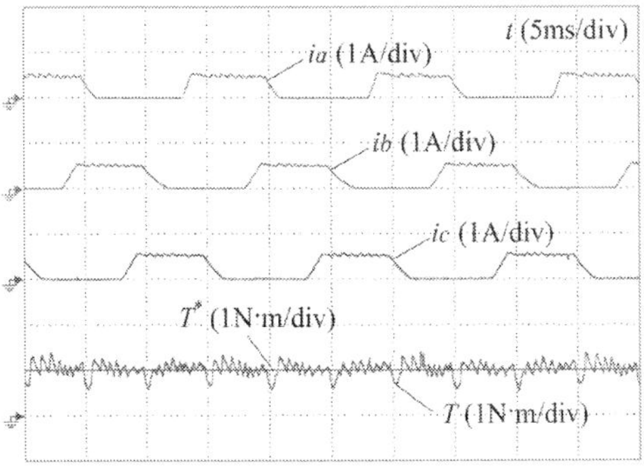

Figure 25. Experimental results of current regulation control mode with fault tolerant topology under fault conditions.

Figure 26 shows the fault tolerance operation at 500 r/min and 5 N·m load in CCC and PWM systems, respectively. The system can still be stable when operating at large load and make up for the missing output torque of the fault phase. Figure 27 shows the operation of the developed system during acceleration and at high speeds with a 1 N·m load. As illustrated in Figure 27(a) , the speed follows the given value well during the continuous acceleration progress. In Figure 27(b) , the system is still stable when it is operated at 1500 r/min, which shows a good stability at high speeds.

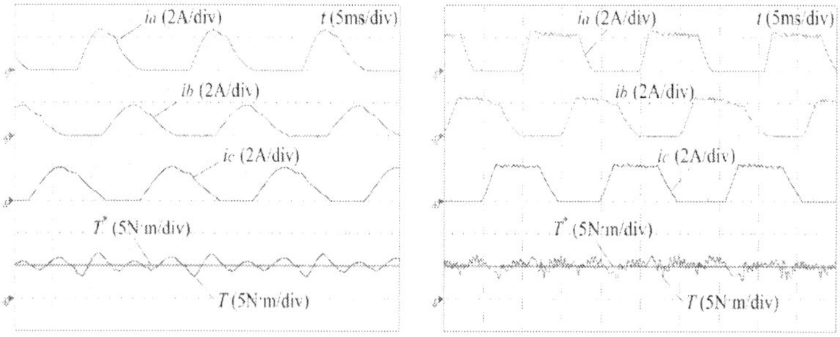

(a) PWM system (b) CCC system

Figure 26. Experimental results of fault tolerance operation under the high load.

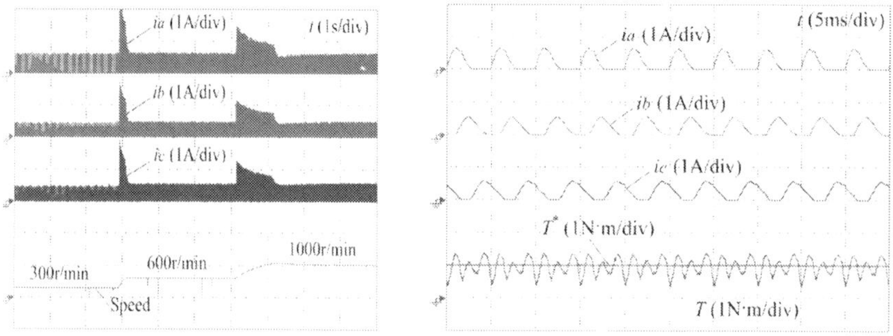

(a) Acceleration (b) High speed operation

Figure 27. Experimental results of fault tolerance operation during acceleration and at high-speed operation.

Figure 28 shows the fault tolerance operation with only a half phase winding at 500 r/min and 1 N·m load in CCC and PWM systems, respectively. The system can still operate with only a half phase winding at light load. Figure 29 shows the fault tolerance operation with only a half phase winding during acceleration and load increasing. The speed still follows the given speed well both during acceleration and in steady state. However, in Figure 29(b) , when the load increases from 1 to 3 N·m, the speed is reduced due to the insufficient load ability. Hence, in the extreme faulty conditions, the proposed fault tolerance scheme can still operate at light loads.

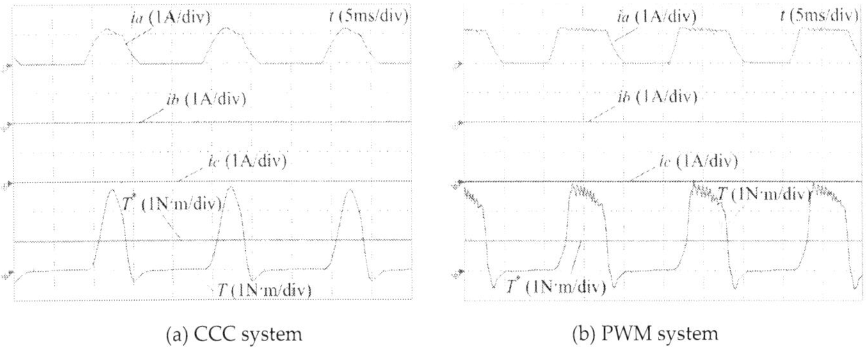

Figure 28. Experimental results of fault tolerance operation with only a half phase winding in the extreme faulty condition at steady-state operation.

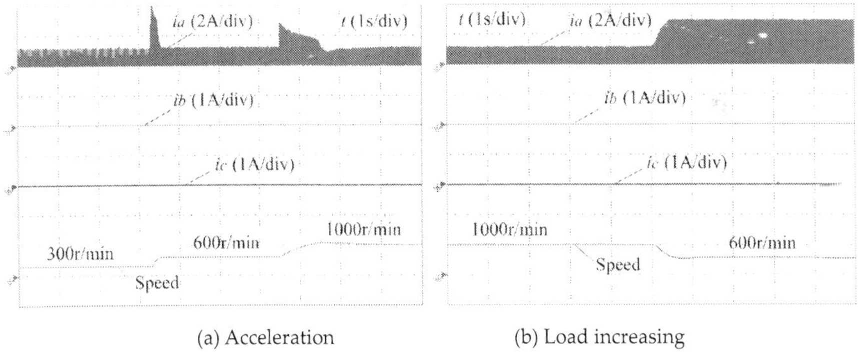

Figure 29. Experimental results of fault tolerance operation with only a half phase winding during acceleration and load increasing.

4. Conclusion

In this chapter, a novel SRM fault diagnosis and fault tolerance strategy have been studied. The main contributions of this work are as follows: (i) The fault characteristics under open-circuit, short-circuit faults are researched on the basis of traditional asymmetrical half-bridge driving topology. (ii) An FFT algorithm with Blackman window interpolation is proposed for SRM fault diagnosis; the method uses the normalized harmonic component of the bus current as the open-circuit fault signature, which can detect the open-circuit faults accurately even in fast transients; the diagnosis time is within one current period, achieving a fast fault detection. (iii) A novel fault tolerance SRM topology is developed with a modular structure. (iv) The developed fault diagnosis technology can achieve convenient fault diagnosis and improve fault tolerance performance of the SRM. Overall, this work will help improve the market acceptance of SRM drives, especially for high-temperature, high-speed, and safety-critical applications.

Author details

Yihua Hu[1], Chun Gan[2], Wenping Cao[3,4*] and Stephen Finney[1]

*Address all correspondence to: w.p.cao@aston.ac.uk

1 Department of Electrical & Electronics Engineering, University of Strathclyde, Glasgow, UK

2 College of Electrical & Electronics Engineering, Zhejiang University, Hangzhou, China

3 School of Engineering and Applied Science, Aston University, Birmingham, UK

4 Department of Electrical Engineering and Computer Science, Massachusetts Institute of Technology, Cambridge, USA

References

[1] C. Onar, J. Kobayashi, and A. Khaligh, "A fully directional universal power electronic interface for EV, HEV, and PHEV applications," *IEEE Trans. Power Electron.*, vol. 28, no. 12, pp. 5489-5498, Dec. 2013.

[2] C. Ruiwu, C. Mi, and C. Ming, "Quantitative comparison of flux-switching permanent-magnet motors with interior permanent magnet motor for EV, HEV, and PHEV applications," *IEEE Trans. Magn.*, vol. 48, no. 8, pp. 2374-2384, Aug. 2012.

[3] J. Hong, H. Lee, and K. Nam, "Charging method for the secondary battery in dual-inverter drive systems for electric vehicles," *IEEE Trans. Power Electron.*, vol. 30, no. 2, pp. 909-921, Feb. 2015.

[4] M. Obata, S. Morimoto, M. Sanada, and Y. Inoue, "Performance of PMASynRM with ferrite magnets for EV/HEV applications considering productivity," *IEEE Trans. Ind. Appl.*, vol. 50, no. 4, pp. 2427-2435, Jul./Aug. 2014.

[5] K. C. Kim, "A novel magnetic flux weakening method of permanent magnet synchronous motor for electric vehicles," *IEEE Trans. Magn.*, vol. 48, no. 11, pp. 4042-4045, Nov. 2012.

[6] S. P. Nikam, V. Rallabandi, and B. G. Fernandes, "A high-torque-density permanent-magnet free motor for in-wheel electric vehicle application," *IEEE Trans. Ind. Appl.*, vol. 48, no. 6, pp. 2287-2295, Nov./Dec. 2012.

[7] S. Morimoto, O. Shohei, Y. Inoue, and M. Sanada, "Experimental evaluation of a rare-earth-free PMASynRM with ferrite magnets for automotive applications," *IEEE Trans. Ind. Electron.*, vol. 61, no. 10, pp. 5749-5756, Oct. 2014.

[8] C. Gan, J. Wu, M. Shen, S. Yang, Y. Hu, and W. Cao, "Investigation of skewing effects on the vibration reduction of three-phase switched reluctance motors," *IEEE Trans. Magn.*, vol. 51, no. 9, Sep. 2015, Art. ID 8203509.

[9] C. Gan, J. Wu, S. Yang, and Y. Hu, "Phase current reconstruction of switched reluctance motors from DC-Link current under double high-frequency pulses injection," *IEEE Trans. Ind. Electron.*, vol. 62, no. 5, pp. 3265-3276, May 2015.

[10] M. Krishnamurthy, C. S. Edrington, A. Emadi, P. Asadi, M. Ehsani, and B. Fahimi, "Making the case for applications of switched reluctance motor technology in automotive products," *IEEE Trans. Power Electron.*, vol. 21, no. 3, pp. 659-675, May 2006.

[11] M. Ehsani, K. M. Rahman, M. D. Bellar, and A. J. Severinsky, "Evaluation of soft switching for EV and HEV motor drives," *IEEE Trans. Ind. Electron.*, vol. 48, no. 1, pp. 82-90, Feb. 2001.

[12] Y. Hu, C. Gan, W. Cao, C. Li, and S. Finney, "Split converter-fed SRM drive for flexible charging in EV/HEV applications," *IEEE Trans. Ind. Electron.*, vol. 62, no. 10, pp. 6085-6095, Oct. 2015.

[13] S. Gopalakrishnan, A. M. Omekanda, and B. Lequesne, "Classification and remediation of electrical faults in the switched reluctance drive," *IEEE Trans. Ind. Appl.*, vol. 42, no. 2, pp. 479-486, Mar./Apr. 2006.

[14] W. Ding, Y. Liu, and Y. Hu, "Performance evaluation of a fault-tolerant decoupled dual-channel switched reluctance motor drive under open-circuits," *IET Electr. Power Appl.*, vol. 8, no. 4, pp. 117-130, Nov. 2014.

[15] N. S. Gameiro and A. J. Marques Cardoso, "A new method for power converter fault diagnosis in SRM drives," *IEEE Trans. Ind. Appl.*, vol. 48, no. 2, pp. 653-662, Mar./Apr. 2012.

[16] J. F. Marques, J. O. Estima, N. S. Gameiro, and A. J. Marques Cardoso, "A new diagnostic technique for real-time diagnosis of power converter faults in switched reluctance motor drives," *IEEE Trans. Ind. Appl.*, vol. 50, no. 3, pp. 1854-1860, May/Jun. 2014.

[17] C. Gan, J. Wu, S. Yang, Y. Hu, W. Cao, and J. Si, "Fault diagnosis scheme for open-circuit faults in switched reluctance motor drives using FFT algorithm with bus current detection," *IET Power Electron.*, in press

[18] D. H. Lee ; J. Liang ; Z. G. Lee ; J. W. Ahn, "A simple nonlinear logical torque sharing function for low-torque ripple SR drive," *IEEE Trans. Industrial Electronics*, vol. 56, no. 8, pp. 3021-3028, Aug. 2009.

[19] V. P. Vujic, "Minimization of torque ripple and copper losses in switched reluctance drive," *IEEE Trans. Power Electronics*, vol. 27, no. 1, pp. 388-399, Jan. 2012.

Chapter 4

Battery Management System for Electric Drive Vehicles – Modeling, State Estimation and Balancing

Jun Xu and Binggang Cao

Additional information is available at the end of the chapter

http://dx.doi.org/10.5772/61609

Abstract

Electric-drive vehicles (EDVs) have drawn more and more attention worldwide. As one of the most important parts of EDVs, battery management systems (BMSs) manage the huge amount of battery cells in EDVs and assures their safety. To achieve these goals, researches in BMS, such as battery modeling, battery state estimation, and battery balancing have gained a new vigor. This chapter will review the existing researches and introduce several advances in these areas.

Keywords: Battery management system, battery model, state of charge, battery balancing, electric-drive vehicle

1. Introduction

Due to pollution and energy crisis, researches in electric-drive vehicles (EDVs), including battery electric vehicle (BEV), hybrid electric vehicle (HEV), and plug-in hybrid electric vehicle (PHEV), have soared worldwide. Considered as the main power of EDVs, battery technology has drawn more and more attention [1-7]. Batteries used in EDVs should not be overcharged or over-discharged to avoid damaging the battery, shortening the battery life, and causing fire or explosions. The battery management system (BMS), with the functions of battery modeling, battery state estimation, battery balancing, etc., is one of the key points to protect the battery and optimize the utilization of the battery in EDVs.

The main purpose of this chapter is to investigate the functions of BMSs for EDVs, researches in which have gained a new vigor. Battery modeling and state estimation are among the most popular topics for BMS, and this chapter will give a detailed introduction to these topics. Battery balancing is also very important in a BMS, considering the "Bucket Effects" for a large amount of battery cells in a battery string. This chapter will analyze the researches in this area according to literature review and a comparative analysis will be given. Several latest research results, including battery modeling, state estimation, and battery balancing, will also be introduced in this chapter.

Firstly, battery modeling methods will be investigated, including the equivalent circuit methods, impedance methods, etc. The comparison between the proposed impedance model and the measured impedance spectra will be shown and analyzed.

Secondly, the state estimation method for the battery used in EDVs, including the state of charge (SOC) estimation, will be analyzed. And a novel adaptive proportional integral observer SOC estimation method will be proposed.

Thirdly, the battery balancing method will be analyzed and compared. The positive balancing methods and active balancing methods will be introduced. A novel balancing topology will be proposed and analyzed.

Fourthly, the experimental validation will be introduced and the results will be analyzed.

Finally, conclusions of this chapter will be given.

2. Battery modeling

As an important interpretation of an actual battery, a battery model could be used as an important reference when designing the vehicle controller and estimating the states of the battery. This section will study battery models describing the characteristics of batteries. Firstly, some existing battery models will be reviewed and analyzed. Then, an impedance battery model will be proposed according to the merits and demerits of the existing battery models.

2.1. Review of battery equivalent circuit models

The battery is an electrochemical system with strong nonlinearity. To model such a strong nonlinear system is very difficult. Some attempts have been made to evaluate the battery models for the state and parameter estimation of a li-ion battery, such as electrochemical models and equivalent circuit models. Electrochemical models take advantage of electrochemical properties of the battery, and this method is relatively too complex for engineering, though it is relatively accurate. Thus, equivalent circuit models is very popular in practical applications, and several equivalent circuit models have been widely used, such as the Rint Model [8-10], the first order RC model [10-12], the second order RC model [10, 13], etc.

2.1.1. The rint model

The Rint model [8] takes advantage of an ideal voltage source V_{oc} as the open circuit voltage (OCV) of the battery. Meanwhile, a resistor R_1 is utilized to interpret the inner resistance of the

http://dx.doi.org/10.5772/61609

battery. $V_{oc}$and R_1 are both functions of SOC, temperature, and so forth and will vary with these parameters. The equivalent circuit of the Rint model is as follows:

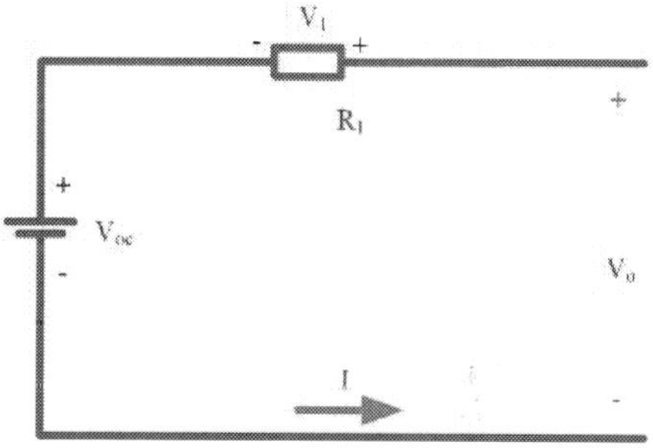

Figure 1. The equivalent circuit for Rint battery model.

In Figure 1, V_1 is used to denote the voltage between terminals of R_1, V_o the voltage of the terminals of the battery, I the current of the battery. In this chapter, the current is assumed to be positive when the battery is charging, while negative when discharging.

According to the circuit theory, following equation could be obtained:

$$V_o = V_{oc} + R_1 I \tag{1}$$

2.1.2. The RC model

The RC model was proposed by the National Renewable Energy Laboratory (NREL), which has already been utilized in the automobile simulation software Advisor [8, 14]. The RC model consists of two capacitors and three resistors. The capacitance of C_c is relatively small to interpret the surface effects of the battery. Meanwhile, the capacitance of C_b is relatively big to interpret the ampere capacity of the battery. The equivalent circuit for the RC model is shown as follows:

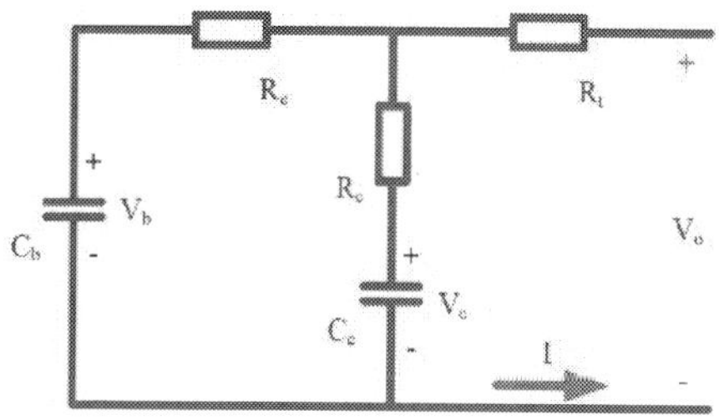

Figure 2. The equivalent circuit for the RC battery model.

Similarly, the mathematical relationship could be obtained as follows:

$$\begin{cases} \dot{V}_b = \dfrac{-1}{C_b(R_e+R_c)}V_b + \dfrac{1}{C_b(R_e+R_c)}V_c + \dfrac{R_c}{C_b(R_e+R_c)}I \\ \dot{V}_c = \dfrac{1}{C_c(R_e+R_c)}V_b + \dfrac{-1}{C_c(R_e+R_c)}V_c + \dfrac{R_e}{C_c(R_e+R_c)}I \end{cases} \tag{2}$$

$$V_o = \frac{R_c}{R_e+R_c}V_b + \frac{R_e}{R_e+R_c}V_c + \left(R_t + \frac{R_eR_c}{R_e+R_c}\right)I \tag{3}$$

where $\dot{V}_b$ and $\dot{V}_c$ means derivation of V_b and V_c respectively.

2.1.3. The first order model

The first order model [9, 15] is realized by adding a parallel RC network to the Rint model, that's why it is called first order model. The added parallel RC network is used to interpret the dynamic response of the battery. The equivalent circuit of the first order model is shown in Figure 3.

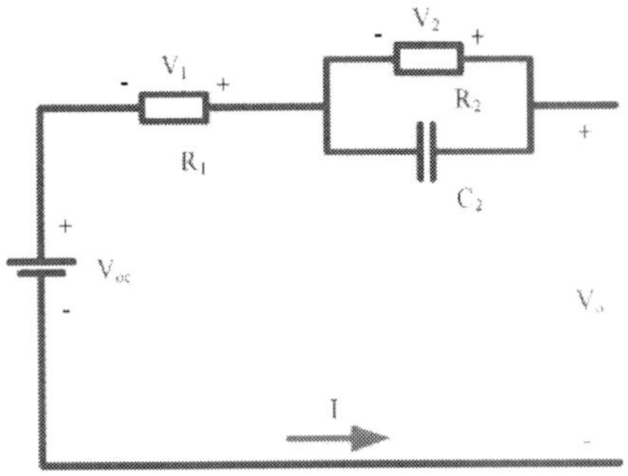

Figure 3. The equivalent circuit for the first order RC battery model.

Four parts could be included in the first order model: the OCV V_{oc}, the inner resistance R_1, the polarization resistance R_2, and the polarization capacitance C_2. V_2 is used to denote the voltage over the parallel RC network. The mathematical relationship is as follows:

$$\begin{cases} \dot{V}_2 = -\dfrac{1}{R_2C_2}V_2 + \dfrac{I}{C_2} \\ V_o = V_{oc} + V_2 + R_1I \end{cases} \tag{4}$$

2.1.4. The second order model

Another parallel RC network added to the first order model forms the second order model. The resistance of the RC network R_3 is used to describe the concentration polarization, while the capacitance C_3 is used to describe the electrochemical polarization. The Second order model is shown in Figure 4 and the mathematical expression is shown in Eq. (5):

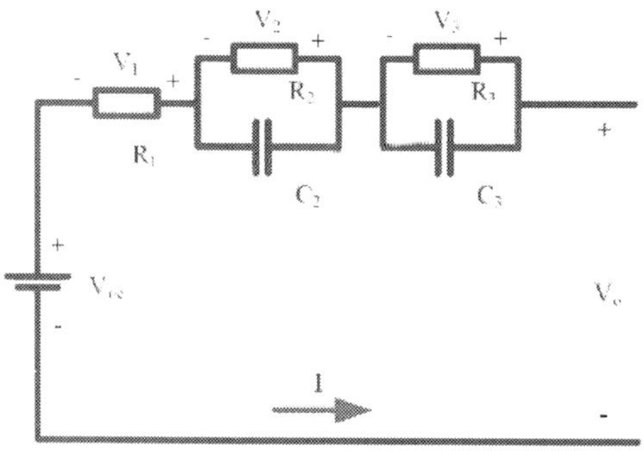

Figure 4. The equivalent circuit for the second order RC battery model.

$$\begin{cases} \dot{V}_2 = -\dfrac{1}{R_2C_2}V_2 + \dfrac{I}{C_2} \\ \dot{V}_3 = -\dfrac{1}{R_3C_3}V_3 + \dfrac{I}{C_3} \\ V_o = V_{oc} + V_2 + V_3 + R_1 I \end{cases} \tag{5}$$

2.2. The battery impedance model

Normally, a more accurate battery model could lead to more accurate state estimation. The models analyzed above have been widely used to estimate the states of the battery, but the problem is that they are sometimes not accurate enough to get satisfying estimation results. The electrochemical impedance spectroscopy (EIS) method is an experimental method to characterize electrochemical systems, and it is considered as one of the most accurate methods to model electrochemical systems, including li-ion batteries and supercapacitors [16-18]. However, studies have shown that the EIS method is too complex to be implemented in real time applications [19]. Besides, the impedance spectra vary with temperature, which adds difficulty to derive SOC directly from the impedance spectra. To solve this problem, an impedance model is proposed in the section [20].

2.2.1. *Impedance spectroscopy test*

In an EIS test, an AC current is applied to the battery, and the voltage response to the current, namely, the amplitude and the phase, is recorded. By the complex division of AC voltage by AC current, the impedance is obtained. These procedures are repeated for different frequencies, and the properties of the battery in full range frequency is obtained [21]. The dynamic behavior of batteries could be activated and recorded by the EIS test, which gives a precise impedance measurement in a wide band of frequencies. The EIS test measures the nonlinearities, as well as very slow dynamics directly, thus it is considered as a unique tool to analyze battery dynamics.

The impedance spectra of a 50% SOC battery with 3 mHz to 2.1 kHz frequency range is shown in Figure 5. Three sections can be obtained for the impedance spectra, as shown in the figure, namely, the low-frequency section, the mid-frequency section, and the high-frequency section [21].

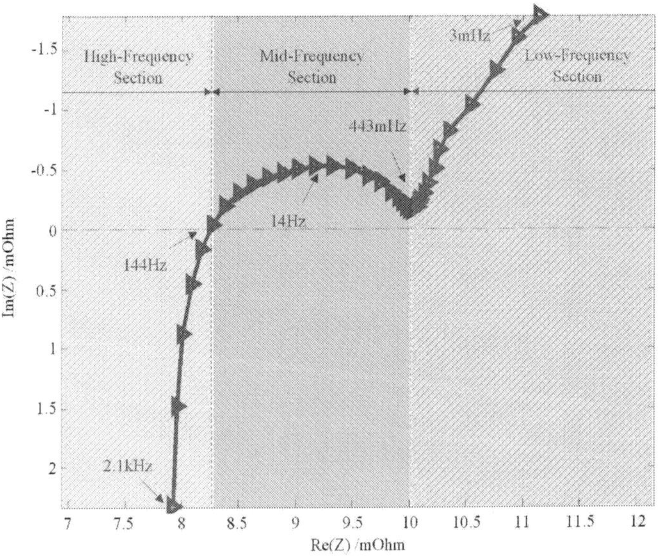

Figure 5. Impedance spectra of a Li-ion battery.

For the high-frequency section, where the frequency is larger than 144 Hz, the impedance spectra intersect with the real axis. According to the electrochemical theories (ET) and circuit theory (CT), this point could be explained by an ohmic resistance in the equivalent circuit.

When the frequency is smaller than 443 mHz, referred to as the low-frequency section, a straight line with a constant slope could be observed. This section could be expressed as a constant phase element (CPE) according to ET [18, 22, 23], which is often referred to as a Warburg element:

http://dx.doi.org/10.5772/61609

$$Z_{Warburg}(j\omega) = \sqrt{\frac{R_D}{j\omega C_D}} \coth\left(\sqrt{R_D j\omega C_D}\right) \tag{6}$$

When in the mid-frequency section (443 mHz–144 Hz), the curve looks like a depressed semicircle. According to ET and CT, this depressed semicircle could be modeled by paralleling a CPE with a resistance in an equivalent circuit, which is called a ZARC element [23]. From the analysis above, the equivalent circuit is depicted as shown in Figure 6.

In Eq. 7, V_{oc} denotes the open circuit voltage of the battery; V_1, V_2, and V_3 denote the voltage for R_1, ZARC, and Warburg respectively; V_o is the voltage output of the battery that can be measured directly from the two terminals of the battery.

$$V_o = V_{oc} + V_1 + V_2 + V_3 \tag{7}$$

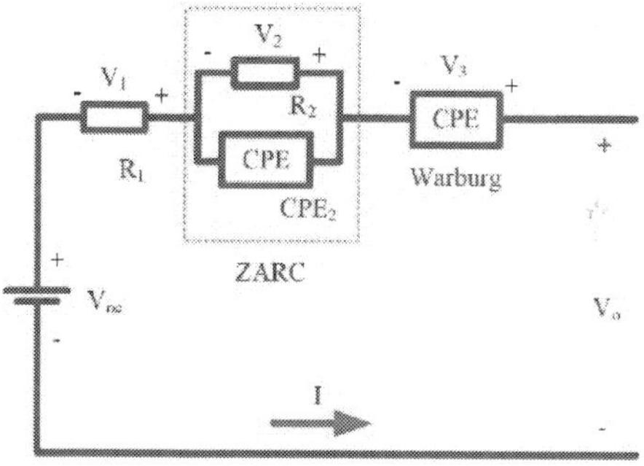

Figure 6. Equivalent circuit for the impedance model.

Then comes the problem: how to use such an impedance model with CPE in it?

2.2.2. *Introduction to fractional order calculus*

Considered as a natural extension of the classical integral order calculus, fractional order calculus (FOC) is becoming more and more popular. According to previous studies, most phenomena, including damp, fluid, viscoelasticity, friction, chaos, dynamic backlash, mechanical vibration, sound diffusion, etc., were considered to have fractional properties [24, 25]. Furthermore, the entire system would have fractional properties, as Machado pointed out, even if parts of it had integral properties [24, 26-28]. Based on FOC, more and more researches have been carried out to develop electrochemical models [22, 29], including lead-acid batteries, supercapacitors, li-ion batteries, fuel cells, and so on. In this section, to represent the impedance model in equations, a fractional modeling method is utilized. By this method, the impedance model can become more convenient to implement the SOC estimation of a battery.

2.2.3. *Impedance model interpreted by FOC*

A FOC modeling method is introduced to interpret such an element. A fractional element is given in the following equation [20, 24]:

$$Z_{fractional}(j\omega) = \frac{1}{Q(j\omega)^r} \tag{8}$$

where $r \in \mathbb{R}$ $(-1 \le r \le 1)$ is the arbitrary order of the fractional element, which can be an integer or a fraction; $Q \in \mathbb{R}$ is the coefficient. When $r = 0$, the fractional element is equivalent to a resistor; when $r = -1$, it is equivalent to an inductor; when $r = 1$, it is equivalent to a capacitor.

Furthermore, the phase of the fractional element listed in Eq. (8) is $r\pi/2$, the magnitude is $20r$ dB dec-1, and the Nyquist plot of the fractional element is a straight line. The slope of the line is a constant value $r\pi/2$, which means the fractional element is a CPE. So, a CPE can be expressed by a fractional element, which is also proved in [24]. In this way, a Warburg element can be modeled by a fractional element $Z_{warburgf}$ as shown in Eq. (9). For the relationship of fractional factor α and the slope of the curve in the low-frequency section, αcan be determined by the curve slope. A Warburg element can be expressed in terms of a fractional element as follows:

$$Z_{Warburg}(j\omega) = \frac{1}{W(j\omega)^\alpha} \tag{9}$$

where $\alpha \in \mathbb{R}$ $(-1 \le \alpha \le 1)$ is an arbitrary number, $W \in \mathbb{R}$is the coefficient.

Similarly, assume the equation of CPE_2 as follows:

$$Z_{CPE_2}(j\omega) = \frac{1}{C_2(j\omega)^\beta} \tag{10}$$

where $\beta \in \mathbb{R}$ $(0 \le \beta \le 1)$ is an arbitrary number, and $C_2 \in \mathbb{R}$ is the coefficient, especially, when $\beta = 1$,

$$Z_{CPE_2}(j\omega)|_{\beta=1} = \frac{1}{C_2(j\omega)} \tag{11}$$

CPE_2 is a capacitor and the capacitance is C_2. The paralleling CPE and resistor will form a semicircle in the Nyquist plot.

Figure 7 depicts the Nyquist plot where $Z_{warburgf}$ is used instead of $Z_{warburg}$ and ZARC is represented by FOC. Figure 7 shows that the fractional method can fit the measured impedance spectra well, which means the FOC impedance model can characterize the battery based on the impedance spectra analysis.

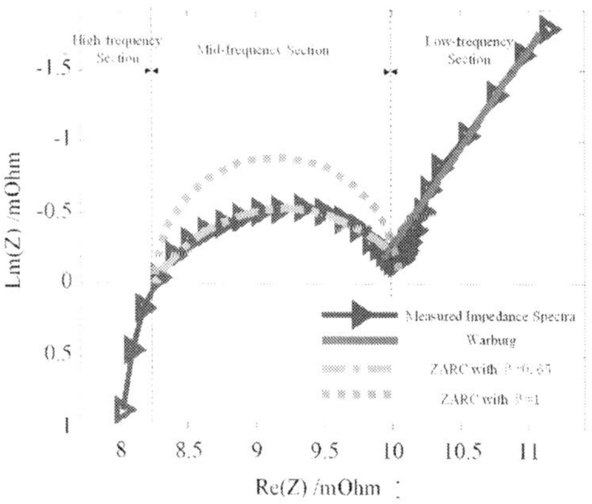

Figure 7. Comparison between measured impedance spectra and the impedance model.

3. State of charge estimation

As an essential indicator for li-ion batteries used in EDVs, SOC is a key state to estimate the driving range of an EDV. Defined as the ratio of the remaining capacity to the nominal capacity of the battery, SOC of a battery can be described as:

$$SOC = \frac{\text{Remaining Capacity}}{\text{Norminal Capacity}} \tag{12}$$

The usable SOC range could be extended if an accurate SOC can be obtained. Thus, a smaller battery pack will be able to satisfy the demand of an EDV that right now is equipped with a large battery pack. The price will be much lower and it will be a great help for market penetration of EDVs.

3.1. Existing methods for state of charge estimation

Many SOC estimation methods have been reported in previous literature, including the ampere-hour method, the OCV method, the model-based method, and so forth. The ampere-hour method is simple and easy to implement for the calculation of battery SOC for it takes advantage of the definition of SOC. However, it needs the prior knowledge of initial SOC and suffers from accumulated errors of noise and measurement error [30, 31]. The OCV method takes advantage of the certain relationship between SOC and SOC, and is considered to be very accurate. However, to obtain the OCV needs a long rest time, and thus it is difficult be used in real-time applications [30]. Intelligent algorithms, such as fuzzy logic, artificial neural networks, and so forth, have been studied to estimate the SOC [32, 33]. Due to the powerful ability to approximate nonlinear functions, these methods can often obtain a good estimation of SOC. However, the learning process for these methods is often quite computationally demanding and complex, which becomes difficult to be applied in online applications.

Model-based SOC estimation methods are the most popular solutions [34-36]. The main methodology is to take advantage of both the voltage and the current of the battery. Measured currents will be applied to the model and the voltages will be calculated using the present and/or past states and parameters of the model. The errors between the calculated voltages and the measured voltages are applied to an algorithm to intelligently update the estimation states of the model. The Luenberger observer [37-39], the Kalman filter [40, 41], and the sliding mode observer [42-44], etc. could be used in such model-based SOC estimation methods.

Proposed by D. Luenberger [45] in 1966, the Luenberger observer is now widely used in different applications. Reference [37-39] introduced it to estimate the battery SOC and owned good results. The Kalman filter uses the entire observed input data and output data to find the minimum mean squared error estimation states of the true states of the li-ion batteries [41]. Essentially, the Kalman filter takes advantage of the prior input currents and output terminal voltages to obtain the Kalman gain. This Kalman gain is like the Luenberger gain, which feedbacks to correct the differences between the calculated states and the true states of the li-ion battery.

Reference [42] introduced the sliding mode observer to estimate the battery SOC. As indicated in the paper, the sliding mode observers inherited the robust properties in SOC estimation. It is robust under modeling uncertainties, but the chatter problem could not be ignored for this method. The block diagram of these methods is shown in Figure 8.

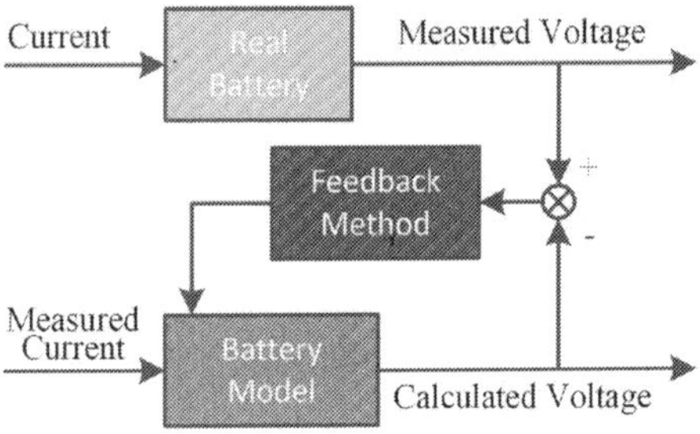

Figure 8. Block diagram of the existing SOC estimation methods.

3.2. The proportional integral observer SOC estimation method

By an additional integrator, the proportional integral (PI) observer is reported to be more robust with respect to modeling uncertainties [46]. Since modeling errors always exist in a battery model, the PI observer is considered to be able to improve the accuracy and estimation speed of SOC estimation. Thus, the PI observer SOC estimation method is proposed in the section. [47]

A battery model could be considered as follows (referred to as System 1):

$$\begin{cases} \dot{x} = Ax + Bu \\ y = Cx + Du \end{cases} \tag{13}$$

However, considering the modeling errors, capacity variation, and so forth, System 1 is not sufficient to model the battery. The nonlinear part should be added to the battery model, which could be described as follows (referred to as System 2) [48-50]:

$$\begin{cases} \dot{x} = Ax + Bu + Ev(x,u,t) \\ \qquad y = Cx + Du \end{cases} \tag{14}$$

where E describes the influence of the nonlinearities to the different states, and such relationships could be obtained by experiments and some "trial and error" approaches; $v(x, u, t)$ describes the nonlinearities, unknown inputs, and un-modeled dynamics of the plant and may be a nonlinear function of states, inputs, and time; $v(x, u, t)$ is referred to as disturbance.

Considering the special applications of the battery for EDVs, the disturbance could be caused by temperature, sensor noise, and so on. Taking temperature as an example, the variation rate could be very slow, and thus $\dot{v} \approx 0$ when the temperature is considered. Meanwhile, the operation temperature range for the battery is limited for the consideration of life cycle and safety. So, $v(x, u, t)$ should also be in a small range due to the influence of temperature. For the strict temperature control in EDV applications, the temperature would be stable after a short time, thus $\lim_{t\to\infty} v(x, u, t)$ exists for the influence of temperature. It is considered to be Gauss Noise with zero mean value for sensor noise. Sensor failure could also be considered to be slow changing, and thus the assumption $\dot{v} \approx 0$ could be reasonable. Actually, since the change rate is so small, the sensor drift could be neglected for a certain drive cycle of the EDV as it changes very little for a one-day drive of the vehicle. It is reasonable to assume that $\lim_{t\to\infty} v(x, u, t)=0$ for the influence of current sensor. So it is assumed that the disturbance $v(x, u, t)$ is not state-dependent, and it is also reasonable to assume that the limitation of disturbance $v(x, u, t)$ exists.

According to the definition of the PI observer, the PI observer is designed as follows:

$$\begin{cases} \dot{\tilde{x}} = A\tilde{x} + Bu + K_p(y - \tilde{y}) + K_{i2}w \\ \qquad \dot{w} = K_{i1}(y - \tilde{y}) \end{cases} \tag{15}$$

Note that variable w is defined as the integral of the difference$(y - \tilde{y})$. Vectors $K_p \in \mathbb{R}^{2\times1}$ and $K_{i1} \in \mathbb{R}^{1\times1}$ $K_{i2} \in \mathbb{R}^{2\times1}$ are the proportional and integral gains, respectively. The design block of the PI observer is given in Figure 9.

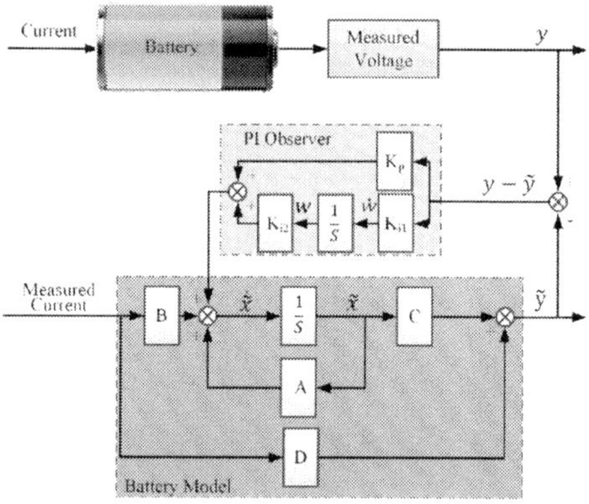

Figure 9. Block diagram of the PI observer SOC estimation method.

The unknown disturbance is considered, which would lead to modeling more accurate battery characteristics. The PI observer is applied to System 2 and when $e_x = \tilde{x} - x$, $e_v = w - v$, and $K_{i2} = E$ are assumed, the error equations could be obtained as follows:

$$\begin{cases} \dot{e}_x = Ae_x - K_p Ce_x + K_{i2} e_v \\ \dot{w} = -K_{i1} Ce_x \end{cases} \tag{16}$$

These equations could be rewritten as:

$$\begin{pmatrix} \dot{e}_x \\ \dot{w} \end{pmatrix} = \begin{bmatrix} A - K_p C & K_{i2} \\ -K_{i1} C & 0 \end{bmatrix} \begin{pmatrix} e_x \\ e_v \end{pmatrix} \tag{17}$$

So,

$$\begin{pmatrix} \dot{e}_x \\ \dot{e}_v \end{pmatrix} = A_e \begin{pmatrix} e_x \\ e_v \end{pmatrix} - \begin{bmatrix} 0 \\ I \end{bmatrix} \dot{v} \tag{18}$$

Since $\dot{v} = 0$ for the certain application as stated above, this equation could be rewritten as follows:

$$\begin{pmatrix} \dot{e}_x \\ \dot{e}_v \end{pmatrix} = A_e \begin{pmatrix} e_x \\ e_v \end{pmatrix} \tag{19}$$

If following matrix pair is observable, A_e could be arbitrarily assigned.

$$\left(\begin{bmatrix} A & E \\ 0 & 0 \end{bmatrix}, \begin{bmatrix} C & 0 \end{bmatrix} \right) \tag{20}$$

which is equivalent to:

$$rank\left\{ \begin{bmatrix} A & K_{i2} \\ C & 0 \end{bmatrix} \right\} = n + r \tag{21}$$

where r is the dimension of v, which is assumed to be 1. Since $n=2$ according to the battery model, the rank should be 3.

Substitute the parameters of the battery model into the matrix:

$$rank\left\{ \begin{bmatrix} A & K_{i2} \\ C & 0 \end{bmatrix} \right\} = rank\left\{ \begin{bmatrix} -\frac{1}{R_2C_2} & 0 & K_{i2_1} \\ 0 & 0 & K_{i2_2} \\ 1 & a_i & 0 \end{bmatrix} \right\} = 3 \tag{22}$$

The matrix is full rank, so the conditions are satisfied that A_e could be arbitrarily assigned.

By utilizing the pole place method or LQ method, $K_p$and K_{i1} could be selected to assure A_e is Hurwitz. Since A_e is Hurwitz, the system is convergent. Thus, when $t \to \infty$, errors will tend to be zero, which means when $t \to \infty$, $\tilde{x}$ would converge to x. Take the battery model in this chapter for example, the estimated SOC would converge to the true SOC.

4. Battery balancing

Battery technology has attracted more and more attention as stated above [51, 52]. However, single batteries could not be used widely in actual applications, since the voltage and capacity of a single battery cell is inherently low. For actual applications, such as utility energy storage in smart grids and EDVs, the battery pack is designed to be hundreds to thousands of volts to optimize the system performance. A large number of battery cells connected in series and parallel to form the battery string are necessary in such applications.

However, no two battery cells could be the same. Such differences could be internal resistance, degradation level, nominal capacity, ambient temperature, etc. These inevitable differences among the cells in a battery string would lead to imbalance, which would cause many problems. Due to the "Bucket Effects", the actual available capacity of the battery string would dramatically decrease. Overcharge and/or overdischarge could also be potentially caused by the imbalance, which is dangerous for batteries. Battery balancing is of paramount importance to maintain the performance and enhance the cycle life.

4.1. Existing balancing topologies

Many battery balancing methods have been presented in previous studies [53-66], which could be categorized as passive balancing methods and active balancing methods. For passive balancing methods [67-69], the excess energy would be removed from high-energy cell(s) through resistors. The energy is dissipated as heat through resistors, which not only wastes electricity energy, but also increases thermal control difficulty of the battery pack. Besides, only the high-energy battery cells work in the balancing process. A large amount of energy would be wasted if most of the battery cells have higher energy than some other cells.

For active balancing methods, active switching circuits are utilized to transfer energy between cells or between the cell(s) and the battery string. Compared to passive balancing methods, wasted energy will be very little. However, to fulfill the functions stated above, more switches and associated components are needed, which would lead to increased cost and reliability concerns. The cost and reliability should be carefully considered to make active balancing more competitive, including but not limited to, reducing the components, simplifying the structure of the topology, and reducing the control complexity.

A bi-directional buck-boost converter-based balancing topology [59-61] is shown in Figure 10a, which is referred to as Topology 1. In this method, a buck-boost converter is formed to transfer energy from a high energy cell to its adjacent cell. The problem is that cells can only be balanced by their adjacent cells. More than one step will be needed if the two cells that need to be balanced have a long distance (C_1 and C_n for instance). In this case, longer time will be needed to transfer energy between these two cells to fullfil the balancing.

Figure 10b shows another widely used balancing topology [62, 63], which is referred to as Topology 2. In this topology, the primary winding of a bi-directional Flyback converter is connected to each cell while the secondary winding is connected to the battery string terminals. A cell to the battery string balancing method is formed in this method, t. cells could be balanced simultaneously since each cell owns a separate converter. The total balancing time could be dramatically reduced compared with Topology 1. However, for n cells battery string, $2n$ switches and $2n$ windings are needed. High voltage stress needs be applied to the switches in the secondary winding of the transformer. The turns-ratio of the transformer should be designed to be large enough to cover the large voltage difference between cells and battery string. These aspects would result in high costs for the system.

Another widely used balancing topology is depicted in Figure 10c [60, 64-66], which is referred to as Topology 3. Only one multi-winding transformer is utilized in this topology. n primary windings of the transformer are connected to n battery cells, respectively, and the secondary winding is connected to the battery string terminals. Only one single converter is utilized in this topology compared with Topology 2. The switches used in this topology are reduced to $n + 1$. Furthermore, high voltage stress will only be applied to one switch in this topology. The secondary windings are also reduced to one. Compared with Topology 2, the cost would be dramatically reduced. However, only one cell could be balanced at a time in this topology, and it is still a string to cell balancing method. Comparing to Topology 2, these aspects lead to a longer balancing time. Besides, the converter efficiency would be almost the same since the transformer turns-ratios are the same as those of Topology 2.

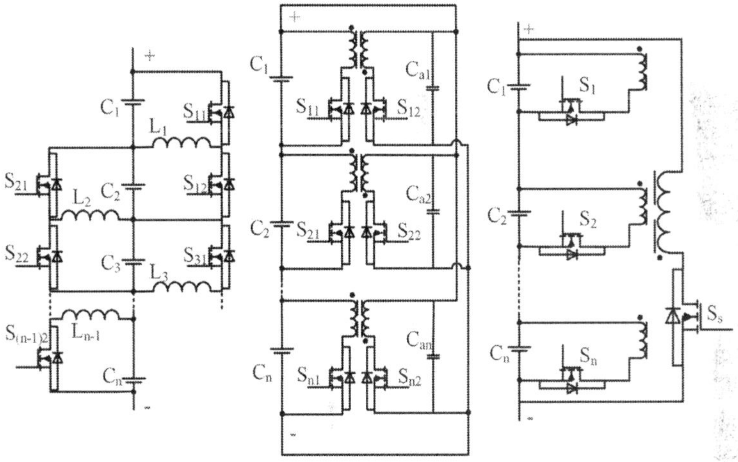

Figure 10. Active balancing topologies: (a) Topology 1, (b) Topology 2, (c) Topology 3.

4.2. The advanced cell to cell battery balancing topology

Based on the analysis of the three popular balancing topologies discussed above (as shown in Figure 11), an advanced cell to cell battery balancing topology is proposed [70]. In this topoloy, the number of switches and the number of windings are reduced. Thus, the cost will be lower and reliability will be higher due to reduced component count. Simultaneously, a much simpler control and gate drive system is needed with fewer switches. Since the topology realizes the cell to cell balancing, balancing time and complexity of the control processes will be reduced a lot. Furthermore, since the transformer turns-ratios are low, almost equals to 1, a relatively higher balancing efficiency and small size transformer could be achieved in this topology.

In Figure 11, L_1, L_2, to L_k are coupled windings, and every two cells in the string share one of such windings. For example, C_1 and C_2 share the same winding L_1. Current I is the main current applied to the battery string, while I_1, I_2,..., and I_{2k} are the balancing current for each cell, respectively. To ensure every two cells sharing a winding, the total number of cells (n) in the battery string are designed to be even ($n = 2k$, k=1, 2...).

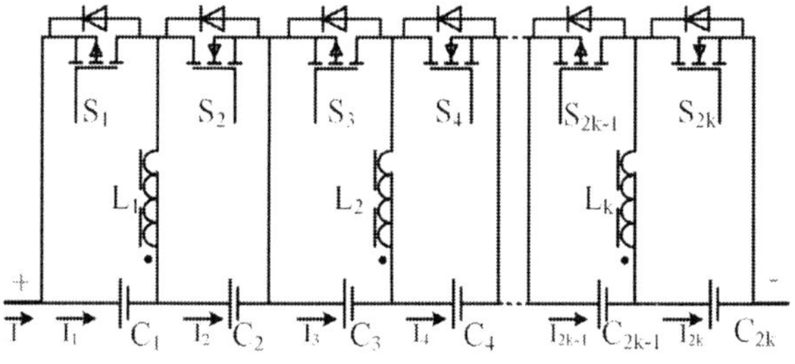

Figure 11. The proposed advanced cell to cell battery balancing topology.

In this topology, the two adjacent cells, which share one winding ($C_1$and C_2 for instance), would form a buck-boost converter similar to that of Topology 1. Meanwhile, due to the coupled windings, non-adjacent cells ($C_1$and C_4 for instance) and its windings form a Flyback converter. So balancing could be easy, no matter if the cells are adjacent or not. The cell to cell balancing method is realized and is considered to be time-saving. To compare this topology with the other above-stated topologies, the components needed for an *n* cell string are compared in Table 1.

	Topology 1	Topology 2	Topology 3	Proposed Topology
Windings	n-1	2n	n+1	n/2
Ferrite Cores	0	n	1	1
Switches	2n-2	2n	n+1	n
Switches Suffering High Voltage Stress	0	n	1	0
Balancing Type	Between Adjacent Cells	Between Cell and String	Between Cell and String	Between Arbitrary Cells

Table 1. Dear authors, please add caption

As shown in Table 1, for the proposed topology, only one switch and one winding are needed for each cell. All the switches used in the proposed topology will not suffer high voltage stress. The cost would be dramatically reduced and the controlling complexity will also be lower.

5. Experimental validation

5.1. Battery test

To learn the characteristics of the battery, different types of battery tests have been taken out in the laboratory.

5.1.1. Capacity test

Capacity test is firstly carried out to obtain the actual capacity of the battery.[71] The testing results are as shown in Figure 12.

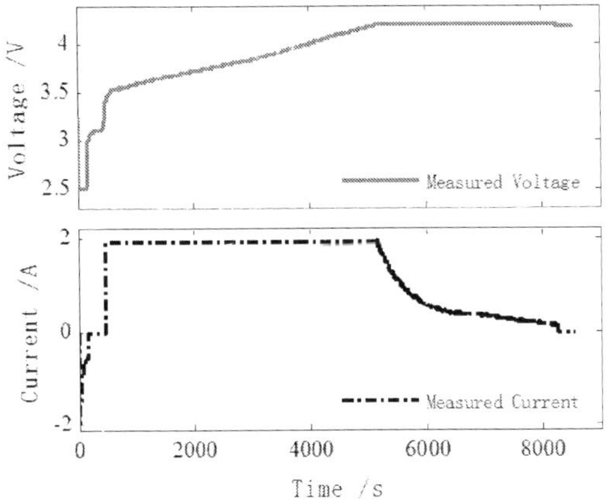

Figure 12. Testing results of the capacity test.

5.1.2. Hybrid pulse power characterization test

The Hybrid Pulse Power Characterization (HPPC) Test is intended to determine the dynamic power capability over the device's useable charge and voltage range using a test profile that incorporates both discharge and regeneration pulses. The results are as follows:

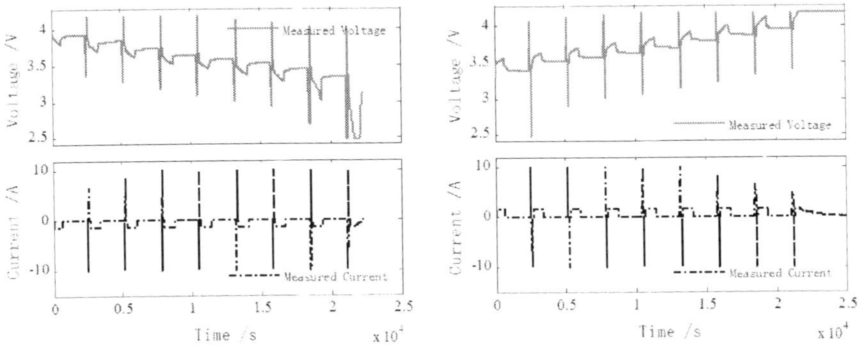

Figure 13. Testing results: (a) HPPC test when discharging; (b) HPPC test when charging.

5.1.3. Driving cycle test

UDDS and US06 drive cycles are applied to the battery. The magnitude of the current profile has been scaled down with respect to the battery features. The current profile of the UDDS drive cycle is given in Figure 14, while the testing results are shown in Figure 15.

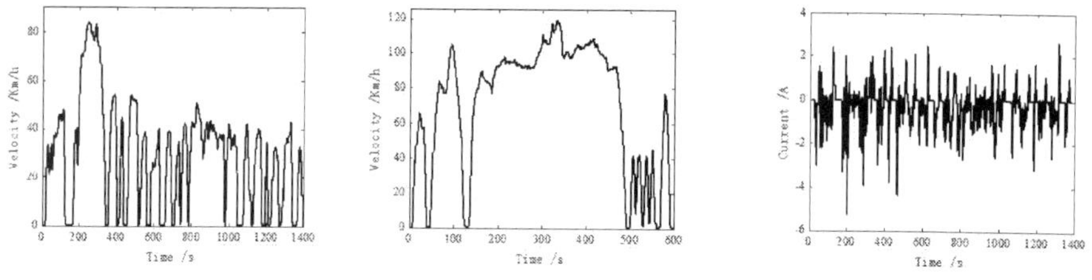

Figure 14. Different driving cycles: (a) UDDS driving cycle; (b) US06 driving cycle; (c) current profile of a UDDS driving cycle.

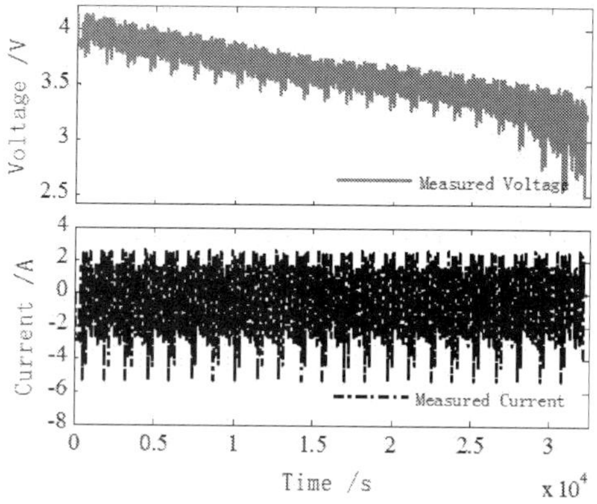

Figure 15. Testing results with UDDS driving cycles.

5.2. Experimental results and analysis

The widely used UDDS current profile is introduced to simulate the drive cycles of an EDV. The ampere counting method is utilized to calculate the reference SOC. The configuration of the battery test workbench is shown as Figure 16.

5.2.1. State estimation validation

Two cases are studied in the validation procedure. The initial SOC is assumed to be given for the SOC estimation method in the first case. The estimation results in this case are given in

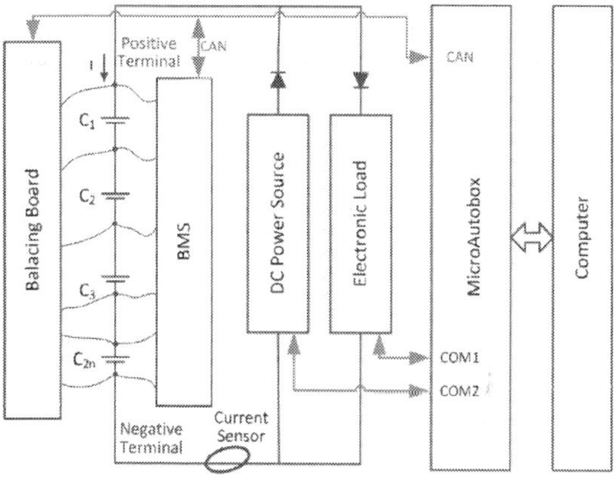

Figure 16. The configuration of the battery test workbench.

Figure 17. The SOC diverges a little in the beginning, since the model is not so accurate. Then, the estimated SOC quickly converges to and keeps on tracing the reference SOC. The results given in Figure 17 show that the SOC estimation method could estimate the SOC with small errors when the initial SOC is given.

The initial SOC is assumed to be unknown in the second case, as shown in Figure 17. In this case, the reference SOC is actually 100% but the initial SOC of the SOC estimation method is assumed to be 60%. It is clear from the figure that the estimation error is 40% at first. Then the estimation errors reduce quickly, and the estimated SOC converges to the reference SOC. After the steady state, the estimated SOC traces the reference SOC with small errors. The results indicate that the initial SOC error could be compensated by the proposed SOC estimation method and the estimated SOC will converge to the reference SOC quickly.

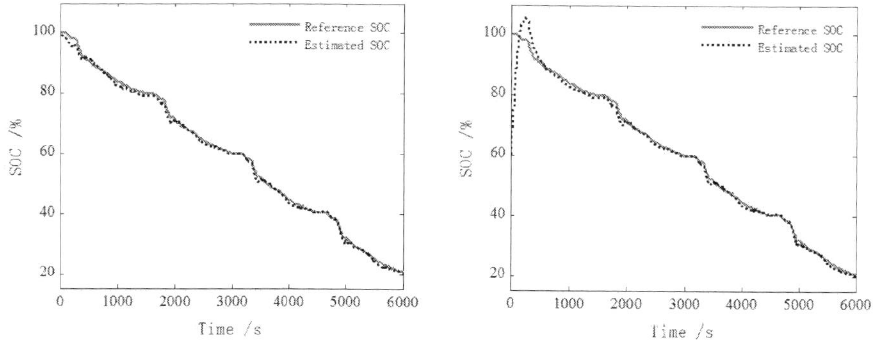

Figure 17. SOC estimation results when the initial SOC is known or unknown.

From the discussions above, it can be concluded that the SOC estimation method works well for li-ion batteries. The SOC estimation errors are small even with a simple battery model, compensating the modeling errors and modeling uncertainties.

5.2.2. Battery balancing validation

To validate the balancing method, the initial SOC is set to 93%, 88%, 88%, and 80% for the four cells, respectively. In this scenario, the energy is balanced from C_1 to C_4. UDDS current profiles will be utilized to test the battery string and the balancing processes are performed according to the SOC-based balancing strategy simultaneously, as shown in Figure 18. The estimated SOC is denoted as rSOC, while the reference SOC is as eSOC in the figure.

Since the balancing process is based on SOC, the SOCs of the four cells should be first known. But these actual SOCs are not known to the SOC estimator. Initially, the SOC estimator assumes that the initial eSOC is 90% for all four cells. However, after a few sampling cycles of 50 to 100 s for example, the eSOC of each cell converges to the rSOC and the SOC differences of the cells can be observer by the balancing algorithm. With the eSOC obtained, the SOC-based balancing algorithms start to work efficiently, as shown in the figure. Eventually, the four cells are balanced according to the SOC in about 4000 s.

With the operation of the balancing procedures (as shown in Figure 18a), the SOC of C_4 and C_1 converge to the average SOC of cells, and finally the SOC of the four cells are almost the same, which could be considered as balanced. The voltage results of the balancing processes are also shown in Figure 18b. The voltages of the four cells are a lot different when the experiment starts, as shown in the first subplot of Figure 18b. After the balancing procedures, the voltages of the four cells converge to almost the same at the end of the experiment as shown in the second subplot of Figure 18b, which is another proof of the effectiveness of the balancing processes.

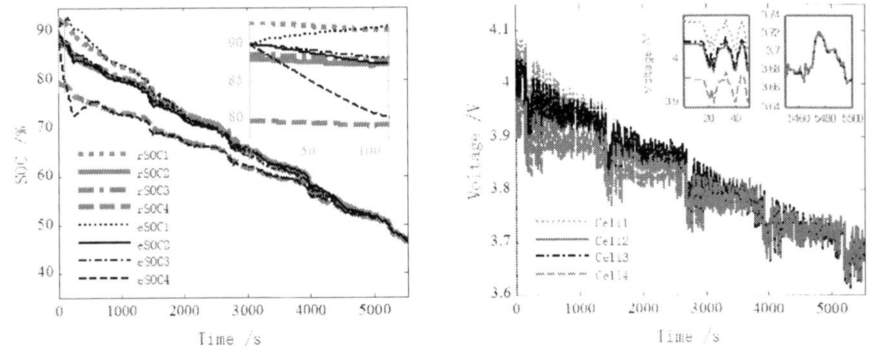

Figure 18. The results of battery balancing experiment: (a) SOC results for the balancing experiment. (b) Voltage results for the balancing experiment.

http://dx.doi.org/10.5772/61609

6. Conclusion

This chapter introduced advances in several popular BMS technologies, including battery modeling, battery state estimation, and battery balancing. Experimental workbench has been established to validate the aforementioned technologies. Several equivalent circuit battery models were first introduced and analyzed. Then based on it, the impedance model and the SOC estimation method were proposed. Several model-based SOC estimation methods were analyzed and the PI observer SOC estimation method was proposed. For the importance of balancing for battery strings in EDV applications, several existing balancing topologies were introduced and a novel cell to cell balancing topology was proposed. Finally, experimental tests were carried out to validate the effectiveness of the methods stated in this chapter. This chapter could be useful for BMS design and further study of technologies in this area.

Acknowledgements

The work was supported by the National Natural Science Foundation of China (51,405,374), the Postdoctoral Science Foundation of China (2014 M560763), the DOE/Chrysler under grant DE-EE0002720, and the China Scholarship Council (CSC) in part. The authors would like to thank Prof. Chris Mi of the University of Michigan, Dearborn for his kindly help.

Author details

Jun Xu* and Binggang Cao

*Address all correspondence to: xujun018@gmail.com

State Key Laboratory for Manufacturing Systems Engineering, School of Mechanical Engineering, Xi'an Jiaotong University, Xi'an, Shaanxi, China

References

[1] X. Wu, G. Guo, J. Xu, and B. Cao, "Application of Parallel Chaos Optimization Algorithm for Plug-in Hybrid Electric Vehicle Design," International Journal of Bifurcation and Chaos, vol. 24, 2014.

[2] X. Wu, B. Cao, X. Li, J. Xu, and X. Ren, "Component sizing optimization of plug-in hybrid electric vehicles," Applied Energy, vol. 88, pp. 799–804, 2010.

[3] B. Wang, J. Xu, B. Cao, and X. Zhou, "A novel multimode hybrid energy storage system and its energy management strategy for electric vehicles," Journal of Power Sources, vol. 281, pp. 432–443, 2015.

[4] B. Wang, J. Xu, and B. Cao, "Design of a novel hybrid power for EV," in Transportation Electrification Asia-Pacific (ITEC Asia-Pacific), 2014 IEEE Conference and Expo, 2014, pp. 1–5.

[5] Z. Chen, C. C. Mi, R. Xiong, J. Xu, and C. You, "Energy management of a power-split plug-in hybrid electric vehicle based on genetic algorithm and quadratic programming," Journal of Power Sources, vol. 248, pp. 416–426, 2014.

[6] Z. Chen, C. C. Mi, Y. Fu, J. Xu, and X. Gong, "Online Battery State of Health Estimation Based on Genetic Algorithm for Electric and Hybrid Vehicle Applications," Journal of Power Sources, vol. 240, pp. 184–192, 2013.

[7] Z. Chen, C. Mi, J. Xu, X. Gong, and C. You, "Energy Management for a Power-Split Plug-in Hybrid Electric Vehicle Based on Dynamic Programming and Neural Networks," Vehicular Technology, IEEE Transactions on, 2013.

[8] G. L. Plett, "Extended Kalman filtering for battery management systems of LiPB-based HEV battery packs: Part 2. Modeling and identification," Journal of Power Sources, vol. 134, pp. 262–276, 2004.

[9] V. H. Johnson, "Battery performance models in ADVISOR," Journal of Power Sources, vol. 110, pp. 321–329, 2002.

[10] H. He, R. Xiong, and J. Fan, "Evaluation of Lithium-Ion Battery Equivalent Circuit Models for State of Charge Estimation by an Experimental Approach," Energies, vol. 4, pp. 582–598, 2011.

[11] K. M. Tsang, L. Sun., and W. L. Chan, "Identification and modeling of Lithium ion battery," Energy Conversion and Management, vol. 51, pp. 2857–2862, 2010.

[12] Y. H. Chiang, W. Y. Sean, and J. C. Ke, "Online estimation of internal resistance and open-circuit voltage of lithium-ion batteries in electric vehicles," Journal of Power Sources, vol. 196, pp. 3921–3932, 2011.

[13] H. He, R. Xiong, X. Zhang, F. Sun., and J. Fan, "State-of-Charge Estimation of the Lithium-Ion Battery Using an Adaptive Extended Kalman Filter Based on an Improved Thevenin Model," Vehicular Technology, IEEE Transactions on, vol. 60, pp. 1461–1469, 2011.

[14] V. H. Johnson, A. A. Pesaran, T. Sack, and S. America, "Temperature-dependent battery models for high-power lithium-ion batteries," presented at the 17th Annual Electric Vehicle Symposium, Montreal, Canada, 2000.

[15] M. Sitterly, W. Le Yi, G. G. Yin, and W. Caisheng, "Enhanced Identification of Battery Models for Real-Time Battery Management," Sustainable Energy, IEEE Transactions on, vol. 2, pp. 300–308, 2011.

[16] P. Mauracher and E. Karden, "Dynamic modeling of lead/acid batteries using impedance spectroscopy for parameter identification," Journal of Power Sources, vol. 67, pp. 69–84, 1997.

[17] S. Buller, E. Karden, D. Kok, and R. W. De Doncker, "Modeling the dynamic behavior of supercapacitors using impedance spectroscopy," Industry Applications, IEEE Transactions on, vol. 38, pp. 1622–1626, 2002.

[18] S. Buller, "Impedance-Based Simulation Models for Energy Storage Devices in Advanced Automotive Power Systems," dissertation, RWTH Aachen University, 2002.

[19] C. Min and G. A. Rincon-Mora, "Accurate electrical battery model capable of predicting runtime and I-V performance," Energy Conversion, IEEE Transactions on, vol. 21, pp. 504–511, 2006.

[20] J. Xu, C. C. Mi, B. Cao, and J. Cao, "A new method to estimate the state of charge of lithium-ion batteries based on the battery impedance model," Journal of Power Sources, vol. 233, pp. 277–284, 2013.

[21] E. Karden, S. Buller, and R. W. De Doncker, "A method for measurement and interpretation of impedance spectra for industrial batteries," Journal of Power Sources, vol. 85, pp. 72–78, 2000.

[22] I. Sadli, M. Urbain, M. Hinaje, J.-P. Martin, S. Raël, and B. Davat, "Contributions of fractional differentiation to the modeling of electric double layer capacitance," Energy Conversion and Management, vol. 51, pp. 2993–2999, 2010.

[23] J. R. McDonald, Impedance Spectroscopy: Emphasizing solid materials and systems: Wiley-Interscience, 1987.

[24] I. Petráš, Fractional-Order Nonlinear Systems: Modeling, Analysis and Simulation: Springer Verlag, 2011.

[25] C. A. Monje, Y. Q. Chen, B. M. Vinagre, D. Xue, and V. Feliu, Fractional-order systems and controls: fundamentals and applications: Springer, 2010.

[26] J. A. T. Machado and A. Galhano, "Fractional Dynamics: A Statistical Perspective," Journal of Computational and Nonlinear Dynamics, vol. 3, pp. 021,201–5, 2008.

[27] J. Xu, J. Cao, B. Cao, and Y. Li, "Fractional three-loop autopilot of air-to-air missile with parameter optimization," Journal of Xi'an Jiaotong University, vol. 45, 2011.

[28] X. Yang, M. Liao, and J. Chen, "A Novel Approach to Processing Fractal Signals Using the Yang-Fourier Transforms," Procedia Engineering, vol. 29, pp. 2950–2954, 2012.

[29] J. Sabatier, M. Cugnet, S. Laruelle, S. Grugeon, B. Sahut, A. Oustaloup, et al., "A fractional order model for lead-acid battery crankability estimation," Communications in Nonlinear Science and Numerical Simulation, vol. 15, pp. 1308–1317, 2010.

[30] B. Pattipati, C. Sankavaram, and K. Pattipati, "System Identification and Estimation Framework for Pivotal Automotive Battery Management System Characteristics," Systems, Man, and Cybernetics, Part C: Applications and Reviews, IEEE Transactions on, vol. 41, pp. 869–884.

[31] K. Kutluay, Y. Cadirci, Y. S. Ozkazanc, and I. Cadirci, "A new online state-of-charge estimation and monitoring system for sealed lead-acid batteries in Telecommunication power supplies," Industrial Electronics, IEEE Transactions on, vol. 52, pp. 1315–1327, 2005.

[32] M. Charkhgard and M. Farrokhi, "State-of-Charge Estimation for Lithium-Ion Batteries Using Neural Networks and EKF," Industrial Electronics, IEEE Transactions on, vol. 57, pp. 4178–4187, 2010.

[33] L. Xu, J. Wang, and Q. Chen, "Kalman filtering state of charge estimation for battery management system based on a stochastic fuzzy neural network battery model," Energy Conversion and Management, vol. 53, pp. 33–39, 2012.

[34] Z. Zou, J. Xu, C. Mi, B. Cao, and Z. Chen, "Evaluation of Model Based State of Charge Estimation Methods for Lithium-Ion Batteries," Energies, vol. 7, pp. 5065–5082, 2014.

[35] J. Xu, B. Cao, Z. Chen, and Z. Zou, "An online state of charge estimation method with reduced prior battery testing information," International Journal of Electrical Power & Energy Systems, vol. 63, pp. 178–184, 2014.

[36] J. Xu, B. Cao, J. Cao, Z. Zou, C. C. Mi, and Z. Chen, "A Comparison Study of the Model Based SOC Estimation Methods for Lithium-Ion Batteries," in Vehicle Power and Propulsion Conference (VPPC), 2013 IEEE, 2013, pp. 1–5.

[37] X. Hu, F. Sun., and Y. Zou, "Estimation of State of Charge of a Lithium-Ion Battery Pack for Electric Vehicles Using an Adaptive Luenberger Observer," Energies, vol. 3, pp. 1586–1603, 2010.

[38] C. Lin, F. Zhang, H. Xue, and X. Lu, "Estimation of battery state of charge using H∞ observer," in Power Electronics and Motion Control Conference (IPEMC), 2012 7th International, 2012, pp. 422–428.

[39] F. Zhang, G. Liu, L. Fang, and H. Wang, "Estimation of Battery State of Charge with H∞ Observer: Applied to a Robot for Inspecting Power Transmission Lines," Industrial Electronics, IEEE Transactions on, pp. 1086–1095, 2012.

[40] G. L. Plett, "Extended Kalman filtering for battery management systems of LiPB-based HEV battery packs: Part 3. State and parameter estimation," Journal of Power Sources, vol. 134, pp. 277–292, 2004.

[41] G. L. Plett, "Extended Kalman filtering for battery management systems of LiPB-based HEV battery packs: Part 1. Background," Journal of Power Sources, vol. 134, pp. 252–261, 2004.

[42] I. S. Kim, "The novel state of charge estimation method for lithium battery using sliding mode observer," Journal of Power Sources, vol. 163, pp. 584–590, 2006.

[43] I. L.-S. Kim, "A Technique for Estimating the State of Health of Lithium Batteries Through a Dual-Sliding-Mode Observer," Power Electronics, IEEE Transactions on, vol. 25, pp. 1013–1022, 2010.

[44] C. R. Gould, C. M. Bingham, D. A. Stone, and P. Bentley, "New Battery Model and State-of-Health Determination Through Subspace Parameter Estimation and State-Observer Techniques," Vehicular Technology, IEEE Transactions on, vol. 58, pp. 3905–3916, 2009.

[45] D. Luenberger, "Observers for multivariable systems," Automatic Control, IEEE Transactions on, vol. 11, pp. 190–197, 1966.

[46] K. K. Busawon and P. Kabore, "Disturbance attenuation using proportional integral observers," International Journal of Control, vol. 74, pp. 618–627, 2012/07/18 2001.

[47] J. Xu, C. Mi, B. Cao, J. Deng, Z. Chen, and S. Li, "The State of Charge Estimation of Lithium-Ion Batteries Based on a Proportional Integral Observer "Vehicular Technology, IEEE Transactions on, vol. 63, pp. 1614–1621, 2014.

[48] C. Jeang-Lin, "Applying discrete-time proportional Integral observers for state and disturbance estimations," Automatic Control, IEEE Transactions on, vol. 51, pp. 814–818, 2006.

[49] D. SÖFFKER, T.-J. YU, and P. C. MÜLLERa, "State estimation of dynamical systems with nonlinearities by using proportional integral observer," International Journal of Systems Science, vol. 26, pp. 1571–1582, 2012/12/20 1995.

[50] S. Beale and B. Shafai, "Robust control system design with a proportional integral observer," International Journal of Control, vol. 50, pp. 97–111, 2012/12/20 1989.

[51] E. Baker, H. Chon, and J. Keisler, "Battery technology for electric and hybrid vehicles: Expert views about prospects for advancement," Technological Forecasting and Social Change, vol. 77, pp. 1139–1146, 2010.

[52] K. C. Divya and J. Østergaard, "Battery energy storage technology for power systems - An overview," Electric Power Systems Research, vol. 79, pp. 511–520, 2009.

[53] L. Yuang-Shung and C. Ming-Wang, "Intelligent control battery equalization for series connected lithium-ion battery strings," Industrial Electronics, IEEE Transactions on, vol. 52, pp. 1297–1307, 2005.

[54] L. Yuang-Shung and C. Guo-Tian, "Quasi-Resonant Zero-Current-Switching Bidirectional Converter for Battery Equalization Applications," Power Electronics, IEEE Transactions on, vol. 21, pp. 1213–1224, 2006.

[55] L. A. Tolbert, P. Fang Zheng, T. Cunnyngham, and J. N. Chiasson, "Charge balance control schemes for cascade multilevel converter in hybrid electric vehicles," Industrial Electronics, IEEE Transactions on, vol. 49, pp. 1058–1064, 2002.

[56] L. Maharjan, S. Inoue, H. Akagi, and J. Asakura, "State-of-Charge (SOC)-Balancing Control of a Battery Energy Storage System Based on a Cascade PWM Converter," Power Electronics, IEEE Transactions on, vol. 24, pp. 1628–1636, 2009.

[57] P. Hong-Sun., K. Chong-Eun, K. Chol-Ho, M. Gun-Woo, and L. Joong-Hui, "A Modularized Charge Equalizer for an HEV Lithium-Ion Battery String," Industrial Electronics, IEEE Transactions on, vol. 56, pp. 1464–1476, 2009.

[58] A. C. Baughman and M. Ferdowsi, "Double-Tiered Switched-Capacitor Battery Charge Equalization Technique," Industrial Electronics, IEEE Transactions on, vol. 55, pp. 2277–2285, 2008.

[59] K. Nishijima, H. Sakamoto, and K. Harada, "A PWM controlled simple and high performance battery balancing system," in Power Electronics Specialists Conference, 2000. PESC 00. 2000 IEEE 31st Annual, 2000, pp. 517–520 vol.1.

[60] X. Aiguo, X. Shaojun, and L. Xiaobao, "Dynamic Voltage Equalization for Series-Connected Ultracapacitors in EV/HEV Applications," Vehicular Technology, IEEE Transactions on, vol. 58, pp. 3981–3987, 2009.

[61] N. H. Kutkut, "A modular nondissipative current diverter for EV battery charge equalization," in Applied Power Electronics Conference and Exposition, 1998. APEC '98. Conference Proceedings 1998., Thirteenth Annual, 1998, pp. 686–690 vol. 2.

[62] C. Karnjanapiboon, K. Jirasereeamornkul, and V. Monyakul, "High efficiency battery management system for serially connected battery string," in Industrial Electronics, 2009. ISIE 2009. IEEE International Symposium on, 2009, pp. 1504–1509.

[63] M. Einhorn, W. Guertlschmid, T. Blochberger, R. Kumpusch, R. Permann, F. Conte, et al., "A Current Equalization Method for Serially Connected Battery Cells Using a Single Power Converter for Each Cell," Vehicular Technology, IEEE Transactions on, vol. PP, pp. 1-1, 2011.

[64] W. Xuezhe and Z. Bing, "The research of vehicle power Li-ion battery pack balancing method," in The Ninth International Conference on Electronic Measurement & Instruments, 2009, pp. 2-498–2-502.

[65] M. Einhorn, W. Roessler, and J. Fleig, "Improved Performance of Serially Connected Li-Ion Batteries With Active Cell Balancing in Electric Vehicles," Vehicular Technology, IEEE Transactions on, vol. 60, pp. 2448–2457, 2011.

[66] W. XueZhe, Z. Xiaopeng, and H. Dai, "The application of flyback DC/DC converter in Li-ion batteries active balancing," in Vehicle Power and Propulsion Conference, 2009. VPPC '09. IEEE, 2009, pp. 1654–1656.

[67] N. H. Kutkut and D. M. Divan, "Dynamic equalization techniques for series battery stacks," in Telecommunications Energy Conference, 1996. INTELEC '96., 18th International, 1996, pp. 514–521.

[68] X. Zhang, P. Liu, and D. Wang, "The Design and Implementation of Smart Battery Management System Balance Technology," Journal of Convergence Information Technology, vol. 6, pp. 108–116, 2011.

[69] Y. Ye, K. W. E. Cheng, and Y. P. B. Yeung, "Zero-Current Switching Switched-Capacitor Zero-Voltage-Gap Automatic Equalization System for Series Battery String," Power Electronics, IEEE Transactions on, vol. 27, pp. 3234–3242.

[70] J. Xu, S. Li, C. Mi, Z. Chen, and B. Cao, "SOC Based Battery Cell Balancing with a Novel Topology and Reduced Component Count," Energies, vol. 6, pp. 2726–2740, 2013.

[71] P. B. T. Manual, "PNGV Battery Test Manual Revision 3," DOE/ID-10,597, February 2001.

Chapter 5

Electric Drives for Propulsion System of Transport Aircraft

Clément Pornet

Additional information is available at the end of the chapter

http://dx.doi.org/10.5772/61506

Abstract

Following the hybridization or complete electrification strategy of the electric drive pursued on terrestrial vehicles, the aviation industry is considering with great attention the application of electrical technology and power electronics for transport aircraft. The growing interest towards electric application in transport aircraft is driven by the ambitious targets for aviation declared by Europe and by the United-States of America and is motivated by the growing evidence that evolutionary improvement of the technologies might not be sufficient to fulfill these targets. As a result, the introduction of disruptive technologies turns out to be essential. With respect to the progress and perspective in electrical technology, electric drive applications to transport aircraft reveals itself to be a promising field in view of meeting the future goals.

The focus of this chapter is the assessment of the electrification of the propulsion system of transport aircraft. While aiming for the ultimate goal of universally-electric aircraft, hybrid-electric approach will be first necessary to match the requirement of aircraft propulsion system and the development pace of the electrical components technology. The combinatorial variety of hybrid-electric and universally-electric propulsion system topology considered for transport aircraft application as well as key enabling technologies are first reviewed. A compendium of hybrid-electric and universally-electric advanced aircraft concepts is then proposed to obtain a notion of the cloud of aircraft configurations and electric drive options investigated up to this point in time. Finally, the integrated prospects of hybrid-electric aircraft are investigated to establish the feasibility of hybrid-electric aircraft for future market segments.

Keywords: Hybrid-electric aircraft, Universally-electric aircraft, Hybrid-electric propulsion topology, Battery-fuel hybrid propulsion system, Aircraft conceptual design, Aircraft Sizing, Aircraft Performance

1. Introduction

Following the hybridization or complete electrification strategy of the electric drive pursued on terrestrial vehicles, the aviation industry is considering with great attention the application of electrical technology and power electronics for transport aircraft. The growing interest towards electric application in transport aircraft is driven first by the ambitious emissions and external noise reduction targets declared by Europe in the Flightpath 2050 program and the corresponding Strategic Research and Innovation Agenda (SRIA) [1] and by the United-States of America in the NASA Environmentally Responsible Aviation N+ series [2]. It is then motivated by the growing evidence that evolutionary improvement of the technologies might not be sufficient to fulfill these targets. Investigations [3] have shown that the reduction potentials might fall short of the 2035 targets and that the deficit becomes even more substantial towards the 2050 goals. This trend can be best explained by the very high maturity reached by contemporary technologies, in particular the propulsion system technology implemented in transport aircraft. Prospects for technological improvements have consequently reached an asymptote leaving not enough potentials for achieving the aggressive targets. As a result, the introduction of disruptive technologies turns out to be essential in view of meeting the future aviation goals. Finally, the progress and perspective in electrical technology development trigger the initiatives and the extensive research activities deployed by the aeronautical community to investigate the feasibility and the potentials of electric technology application to transport aircraft.

A current application of electrical technology for transport aircraft is the so-called more-electric aircraft [67-71]. The objective of the more-electric aircraft initiative, which targets the aircraft power systems, is basically to replace pneumatic and hydraulic systems by electrical systems. The Boeing B787 is the first aircraft utilizing a more electric power system architecture. A logical future conceivable step is the electrification of the propulsion system of the aircraft which is the topic of this chapter. While aiming for the ultimate goal of an universally-electric aircraft, hybrid-electric approach will be first necessary to match the requirement of aircraft propulsion system and the development pace of the electrical components technology. Hybrid-electric aircraft feature typically a combined conventional and electrical propulsion system. The combinatorial variety of hybrid-electric and universally-electric propulsion system topology considered for transport aircraft application as well as enabling technologies are first discussed in Section 2. A compendium of hybrid and universally-electric advanced aircraft concepts is then proposed in Section 3 to obtain a notion of the cloud of aircraft configurations and electric drive options investigated up to this point in time. The feasibility of hybrid-electric aircraft needs to be established for future market segments. On the basis of selected concepts, the integrated prospects of hybrid-electric aircraft are finally investigated in Section 4.

2. Hybrid-electric and universally-electric propulsion system architecture

An overview of the topological variety of propulsion systems featuring an electric drive approach is presented in this section. Electric approaches to aircraft propulsion system are known as hybrid-electric and universally-electric [4]. By distinguishing between the generation and the transmission to the drive-shaft, the components and their possible combination for the layout of hybrid-electric and universally-electric propulsion system architecture can be illustrated as in Figure 1 [5]. Focusing on the propulsion system, the consumer of electric power is the propulsive device. Ducted and unducted propulsors are typically considered to provide the thrust required to propel transport aircraft. The commonly known-term ducted-fan is understood to be covered by the category ducted propulsor while unducted propulsor includes propeller and open-rotor arrangements.

In a conventional propulsion system, shaft power is generated by burning fuel in a gas-turbine. A gas-turbine converts chemical energy into mechanical power through an aero-thermodynamical process. In an electrical propulsion system, shaft power is produced through a process that can be subdivided into the generation, the distribution and the conversion of electrical energy into mechanical power.

For transport aircraft application, batteries and fuel cells are considered for the generation of electrical energy. Even if not explicitly shown in Figure 1, supercapacitors or flywheels could be also considered in a layout of a hybrid-electric propulsion system [6]. Another means of producing electrical power is through the utilization of a generator driven by a conventional gas-turbine. Known under the terminology of turboelectric, it is discussed in more detail in Section 2.1.1. The electric energy is monitored and distributed from the sources to the consumers by a Power Management and Distribution system (PMAD). The PMAD is typically composed of controllers, converters, inverters, cables, electrical buses and circuit breakers. The layout of the PMAD system is a rather complex task. Optimum layout of the PMAD system results from the analysis of system efficiency, mass, bill-of-material, reliability and maintenance under constraints of abnormal mode of operation. Another important aspect related to the electrical system is thermal regulation to provide for every operating condition an appropriate thermal environment to the electrical and power electronics components. The consideration of thermal management is essential in the early design phase of the integrated electric drive due to its highly-interlaced interactions with the propulsion and power systems. Presenting the development of a framework for concurrent sizing of the powertrain and the thermal management system, Freeman et al. [7] highlighted the design options and implications of thermal management. Three main options for thermal regulation were discussed [7] including air cooling, liquid cooling and cryogenic cooling system providing the advantages and drawbacks of each options in terms of system performance, integration implications, complexity and bill-of-material. The critical importance of adequate flow rate for heat transfer ensuring a proper thermal regulation of the electrical components was highlighted. Freeman et al. [7] predicted the revival of radiator technology due to the introduction of electric drive. The design considerations of radiators were described as well as their interactions with the propulsion system and the implications at aircraft level in terms of weight and parasitic drag. The potential synergistic use of the excess heat produced by the electrical components for anti-icing system, cabin Environmental Control System and galleys were also indicated [7]. The thermal management system needs to be integrated globally. The necessity of a transverse approach is argued by Liscouet-Hanke [8] in order to avoid developing localized dedicated engineered thermal management solutions for each of the electrical components.

In order to drive the propulsor device, the electric energy is converted into shaft power by an electric motor (see Section 4.2). It can drive by itself the propulsor device or it can be mounted on the shaft of a combustion engine to support its operations. This latest arrangement known as a parallel system is discussed in Section 2.1.2. Due to the nature of the electric energy which can be easily partitioned, the field of distributed propulsion which aims to achieve highly coupled structural-aero-propulsive configurations, is often combined to hybrid-electric and universally-electric approaches (see Section 2.2.3)

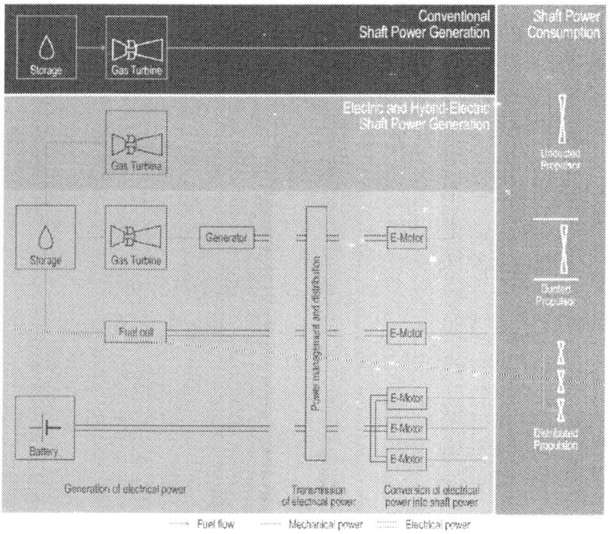

Figure 1. Conventional, hybrid and universally-electric propulsion system topology [5]

2.1. Topological options for electric drive application

The topological options for the application of electric drives to the propulsion system of transport aircraft are detailed in the following sections. The discussion covers serial, parallel as well as integrated arrangements.

2.1.1. Serial system

A general characterization of a serial system is given by the electrical nature of the node connecting the different systems composing the propulsion power-train. The most common serial arrangement is known as turboelectric [9–12]. It denotes a serial system where electric power is produced by a generator driven by a combustion engine typically a gas-turbine. Efficiency improvement in the propulsion system results notably from the advantage that the gas-turbine operation is now decoupled from the operational constraints of the propulsor [9, 13]. The system efficiency and mass can consequently be optimized by operating the gas-turbine and the propulsor close to their peak efficiency. However, because of the greater bill-of-material due to the additional electrical components, in the propulsion chain, the weight of the system is expected to increase compared to a conventional propulsion system [9]. In order to make this approach viable at aircraft level, this penalizing aspect needs to be overcome by any improvements in system efficiency and/or structural-aero-propulsive integration (see Section 2.2.3).

A serial system using fuel-cells as means of electrical energy generation can be entertained. The efficiency of advanced fuel cells including the balance-of-plant is forecast to reach efficiency levels comparable to that of advanced gas-turbines. However, the power density of fuel cell stack is expected to remain much lower than that of a gas-turbine [14]. The

http://dx.doi.org/10.5772/61506

weight penalty might consequently outweigh potential benefits. The unique utilization of fuel cells for providing the power requirement for the propulsion of transport aircraft remains consequently challenging [15].

A serial system whose electrical energy providers are batteries solely was dubbed as universally-electric architecture [4]. The clear advantage of utilizing battery in a propulsion system is the efficiency. With values remaining above 90% during a complete mission profile [16] significant improvement in overall propulsion system efficiency can be achieved compared to a conventional system. However, the gravimetric specific energy (the amount of energy content per unit mass) of advanced batteries is expected to remain relatively low, with a factor of about 8 compared to fossil fuel based on a complete system exergy analysis [17]. Consequently, the weight of the battery and its detrimental sizing cascading impacts on the overall aircraft penalizes the efficiency benefit.

As a result of all these considerations, serial arrangement combining for instance a turboelectric system with battery and/or fuel cell system are considered [14] to draw the advantages of each of the system and to create system synergies in order to achieve greater system performance to the detriment of a higher system complexity. This is notably the approach undertaken by Airbus with the E-Thrust concept [18], which combines a turboelectric system with batteries. Innovative, synergistic integrated serial systems need to be further investigated at aircraft level in order to assess the full potential of hybrid-electric serial propulsion system.

2.1.2. Parallel system

A parallel system is characterized by mechanical nodes that connect the different systems. The most common parallel approach is the installation of an electric motor on the low-pressure shaft of a gas-turbine in order to support the operations of the gas-turbine or even drive by itself the propulsor device during segments of the mission [19–22]. Because of the benefit of utilizing battery on the overall propulsion system efficiency, the electric motor is commonly powered by batteries but the utilization of fuel cells is also conceivable. However, it was found that driving simultaneously the shaft of the gas-turbine by an electric motor influences dramatically the operation of the gas-turbine. The simultaneous operation of the electric motor forced notably the gas-turbine to operate into part-load impairing its efficiency. Moreover, due to the modification of the operating line of the gas-turbine components, the margin to surge might also become critical. Practical engineering solutions need to be envisioned for the integration of the electric motor in the environment of the gas-turbine. Parallel integration of an electric motor on the low-pressure shaft could consequently disrupt contemporary design axioms of gas-turbines.

2.1.3. Distributed parallel system

In view of these challenges and motivated by the search for more synergistic integration, an innovative approach to a parallel hybrid-electric propulsion system was proposed by Pornet and Isikveren [23] taking advantage of distributed propulsion technology (see Section 2.2.3). Instead of coupling the electric motor to the shaft of the gas-turbine, the electric motor is coupled directly to the shaft of the propulsor and the combination electric motor and propulsive device (called electric-fan) is integrated on the aircraft as an additional bill-of-material item to the conventional combustion based engines. Concrete aircraft concepts would be a tri-fan aircraft with two fans conventionally powered by gas-turbines while the remaining fan is driven by an electric motor (or vice versa with two electric fans and one turbofan aft-fuselage mounted), or, as a quad-fan aircraft equipped with two turbofans and two electric fans. The integrated prospects of this latest concept are the subject of the investigation presented in Section 4.4. This approach offers numerous advantages compared to mounting the electric motor on the low-pressure shaft of the gas-turbine. As the conventional system is decoupled from the electrical system, design and operation of the conventional and electrical system are independent. As a result, contemporary design and off-design heuristics of the gas-turbine are not perturbed by the introduction of the electrical system. Moreover, it reduces the system complexity and clears out the integration challenges of the electric motor in the environment of a gas-turbine. By thoughtful sizing and operational strategy of the gas-turbine and the electric motor, the efficiency of the hybrid propulsion system can be optimized by running the conventional and electrical system close to their peak efficiencies. This innovative parallel hybrid arrangement marries up perfectly with distributed propulsion technology opening potentials for tightly aero-propulsive-structural integration.

While in the architecture investigated in [23], the electric motors are powered solely by batteries, a further topological evolution of the distributed propulsion system can be conceived with the introduction of a turboelectric approach. By equipping the gas-turbine with a generator, additional electric power can be transmitted to the electric motors. This system approach reduces the technological level requirement imposed to the battery in terms of gravimetric specific energy while enabling significant increase in system efficiency when using the highly efficient battery system for propulsion. This topology is also interesting as it enables the possible combination of charge sustaining and charge depleting strategies of the batteries for optimum energy management [24, 25]. Charge sustaining strategy, recharging the battery with the generator utilizing excess power of the gas-turbine during segments of the mission, would reduce the integrated battery pack mass and volume requirement. A schematic representation of this topology is given in Figure 2.

2.1.4. Integrated system

While not illustrated in Figure 1, another recent approach for hybrid-electric system to be considered is the so-called integrated system [26, 27] which consists of electrifying part of the core cycle of a gas-turbine. A possible configuration for hybrid-electric integrated system was proposed by Schmitz and Hornung [28] investigating the electrification of the high-pressure compressor stages of a gas-turbine. Still in a pioneering phase, few publications are currently available on this topic but it is definitely an application to follow closely as it gathers momentum.

2.2. Enabling technologies for electric drive application to transport aircraft

Technologies considered as key enablers for the deployment of electric drives to the propulsion system of transport aircraft are discussed in this section. The application of the field of superconductivity to electric propulsion system is first reviewed followed by a discussion about distributed propulsion technology.

2.2.1. Energy storage

Technology improvement of the electrical components discussed in Section 2 in terms of gravimetric specific power and efficiency are of particular importance for the application

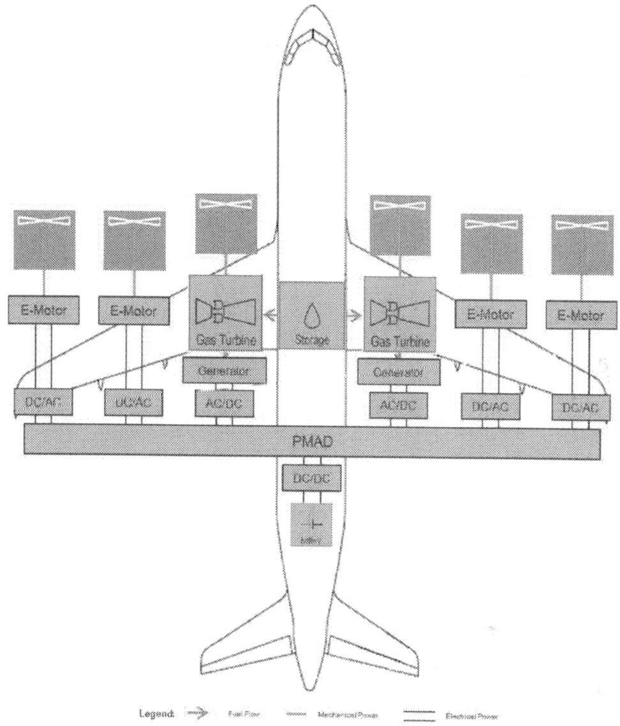

Figure 2. Distributed parallel hybrid-electric topology for distributed fans arrangement

of electric drive to transport aircraft. But most importantly, the evolution of energy storage capability, in particular, the progress made in battery technology is of great relevance for establishing the competitiveness of hybrid-electric and universally-electric aircraft. Gravimetric specific power (the amount of power delivered per unit of mass) and gravimetric specific energy (amount of energy contained per unit of mass) are the key metrics for evaluating the battery application. These parameters are not independent characteristics and they are determined during battery design [29]. As a result, the combined enhancement of the specific energy and the specific power turns out to be a greater technological challenge than the single criterion optimization. Additional aspects such as gravimetric specific energy density, efficiency, design service goal, operating temperature, discharge behavior, safety, reliability and environmental sustainability need also to be considered when selecting the battery technology [30].

Lithium battery technology is regarded as a promising option for application in aviation. State-of-the-art lithium batteries reach specific energy of circa 200 Wh/kg at cell-level [31]. Extensive research focusing on improving the cathode and anode materials as well as on enhancing electrolyte properties [30, 32] aim to scale up the power and energy capability of batteries for aircraft application. Specific energy of around 400 Wh/kg is predicted to be achieved at cell-level through future progress in lithium battery technology [33].

Breakthrough in battery technology could be achieved through the development of innovative battery concept as indicated by open battery systems like zinc-air, aluminium-air and lithium-air [32, 34]. Lithium-air batteries are considered with attention for aircraft application with estimated theoretical gravimetric specific energy from 1000 Wh/kg [33, 35] up to 2000 Wh/kg [36] at cell-level. Not currently commercially available, market readiness of lithium-air is expected for a time-line horizon of 2030 [37].

2.2.2. *Superconductivity*

Superconductivity is considered as an enabling technology for electric drive application to transport aircraft [13, 38, 39]. Through the utilization of High-Temperature Superconducting (HTS) materials, the gravimetric specific power and the efficiency of the electrical component can be significantly improved. The most common application of HTS materials at this point in time is for electric motors and generators [38, 40, 41]. The application of HTS technology to the transmission cable is also considered [10, 11]. The necessity of developing fully superconducting network (including the fault management, protection and switching implications) was argued by Malkin and Pagonis [42]. The challenge of HTS application for transport aircraft lies essentially on the requirement to operate at cryogenic temperature and the resulting complex integration of the cooling system. Instead of using fossil fuels to operate the aircraft, the utilization of cryogenic fuel such as liquid hydrogen can result in strong synergies with the layout of HTS electrical system as the coolant is already available [10, 13, 39]. Handling safety related issues [43], the negative integration impacts of the cryogenic tank on aircraft design [44] and the infrastructural challenges to supply the liquid hydrogen to the operated airports all are issues that remain unresolved [45, 46].

2.2.3. *Distributed propulsion*

The investigation of hybrid-electric or universally-electric system is often coupled with distributed propulsion technology [72-75]. This combination is explained by the nature of the electric energy, which can be easily distributed and by the search for aero-structural benefits through higher integration of the propulsive device with the airframe. The field of Boundary Layer Ingestion (BLI) [47, 75] consisting of re-energizing the low momentum boundary layer in view of aerodynamic efficiency improvement, becomes central. This is mainly the reason why distributed propulsion has been intensively investigated on Blended Wing Body (BWB) configuration [12, 48] as it offers large potential for application of BLI by distributing buried propulsion devices along the trailing-edge of the fuselage (see Section 3). For BLI application on tube and wing configurations, the Propulsive Fuselage configuration, which is characterized by a large fan encircling the rear end of the fuselage, was evaluated as most promising and was the center of several investigations [49–53].

Electric distributed propulsion technology is expected to disrupt the traditional aircraft design paradigms [54–56]. A salient example is the redesign of the wing for optimum efficiency in cruise enabled by distributing propellers along the leading-edge as initiated in the LeapTech Project [57] (see Section 3). Usually, wing design is constrained by low-speed operations in order to achieve according to the properties of the high-lift system acceptable takeoff and landing field performance. Benefiting from the propeller slip-stream effect on the wing, the low-speed requirement on high-lift devices and wing design could be reduced, opening the design space for optimum wing design for en-route operations. Moreover differential thrust could be applied to control the aircraft reducing the requirement on flight control surfaces. The distribution of thrust along the wingspan could enable for instance the control of the aircraft yawing motion resulting in reducing or even conceived the complete removal of the rudder. The one-engine-inoperative case is a very stringent low-speed condition for the sizing of the propulsion system, vertical stabilizer and flight control surfaces in order to comply to the airworthiness regulations and the aircraft top-level field performance requirement. Because of the intrinsic redundancy in thrust production provided by distributed propulsion technology and through a proper redundancy definition in the energy and power system, failure modes either energy/power system inoperative or propulsive device inoperative would result in less severe penalty on system sizing [58, 59]. It is also conceivable that the airworthiness regulations notably with respect to the climb gradient requirements will need to be revisited to adapt to the characteristics of aircraft employing distributed propulsion technology. These highlighted potential benefits in aircraft design demonstrate that the full-benefit of hybrid-electric and universally-electric propulsion system can only be assessed through a holistic integration at aircraft level.

3. Compendium of hybrid- and universally-electric aircraft

While the first fixed-wing electric flight took place for over 9 min in 1973 with the Brditschka's MB-E124, it took around 30 years before reconsidering seriously electric propulsion system for transport aircraft application. This time lapse observed in the interest revival for electric drive application to aircraft propulsion is explained first by the time required for electrical component technology to evolve to a level applicable to the propulsion system requirement of transport aircraft. Secondly, it is driven by the growing environmental awareness and the search for an alternative to a fossil fuel economy. The late 1990 witnessed the birth of several electric experimental and commercial aircraft targeting the general aviation sector in the one or two seater category. While mainly motivated by engineering curiosity through the impulse of pioneers, the industry is currently demonstrating a strong growing interest in the development of hybrid-electric and universally-electric aircraft. A compendium of experimental, commercial as well as advanced hybrid-electric and universally-electric aircraft concepts, was proposed by Pornet and Isikveren [23]. By illustrating the cloud of concepts in Figure 3 evaluating the number of passenger (PAX) versus the design range (evaluated in nautical miles [nm]) certain clusters and trends in design can be identified.

This compendium comprises universally-electric aircraft concepts using battery as energy source with the four seater Airbus E-Fan [18], the four seater LeapTech concept [56, 57], the Dornier Do328-LBM [60] , the Voltair concept [61] and the BHL Ce-Liner [4] as well as hybrid-electric aircraft, integrating a battery-fuel system either in serial or parallel topology, with the NXG-50 [62], the Sugar-Volt [19], the Bauhaus Luftfahrt Twin-Fan [22] (see Section 4.3) and the Bauhaus Luftfahrt Quad-Fan [23] (see Section 4.4). On the upper right corner, BWB configuration using turboelectric distributed approach are represented with the BW-11 [48] and the N3-X [12].

As highlighted by [54], electric propulsion technology will emerge first in general aviation as it provides benefit advantages for early market success and will evolve with respect to the maturation and the development of electric technology towards application for commuter, regional and narrow-body transport aircraft. A noticeable design implication of utilizing

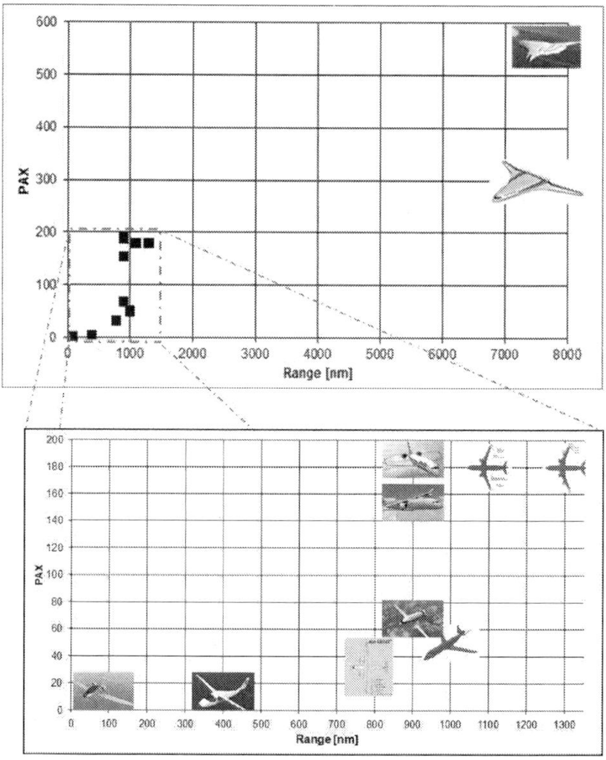

Figure 3. Compendium of advanced hybrid- and universally-electric aircraft. Adapted from [23]

an electric drive relying on battery technology is the achievable design range as indicated by the low range end on the left corner of the chart which reflects more of the regional market segment. With regards to morphology, besides the BWB configuration, no large departure from the traditional "tube and wing" were foreseen due to the implementation of hybrid-electric propulsion system. Outcome of pre-design studies investigated by Isikveren et al. [63] indicated that unless significant departure in the propulsion system integration is entertained, considering for instance distributed propulsion technology, the contemporary tube-and-wing morphology was still considered to be appropriate. Distributed propulsion technology could considerably disrupt contemporary aircraft design paradigm. This was exemplified by the LeapTech concept through distribution of multiple propellers on the wing leading-edge (see Section 2.2.3) and the Propulsive Fuselage configuration selected for the Voltair concept (see Section 2.2.3). The aviation community is in the midst of a pioneering era with regards to electro-mobility and faces an explosion of combinatorial possibilities at system level and aircraft level. Innovative approaches to electrically driven propulsion system needs to be further analyzed and thoughtfully integrated at aircraft level to determine the full potentials. With growing understanding of the technology and its implication at aircraft level, innovative advanced aircraft configurations designed through more ambitious holistically integrated electric propulsion system are expected to emerge.

http://dx.doi.org/10.5772/61506

4. Insights regarding integrated hybrid-electric transport aircraft

By reviewing concepts out of the cloud illustrated in Figure 3, insights regarding the integrated prospects of hybrid-electric transport aircraft are presented in this section. The development of algebraic parameters and figure-of-merits to enable the assessment of aircraft utilizing electric drive for the propulsion system are introduced first. The integration implications of the hybrid-electric propulsion system on aircraft overall sizing and performance are then detailed based on three investigated hybrid-electric narrow-body transport aircraft concepts.

4.1. Algebraic descriptors and figure-of-merits

The algebraic description of a hybrid-electric propulsion requires the establishment of two parametric descriptors [63, 64]: the Degree of Hybridization for Power (Hp) and the Degree of Hybridization for Energy (He). The parameter Hp describes the amount of electrical power relative to the overall total power. Commonly, the installed power or the useful power (power measured at the propulsor) is quoted. In the analyzes provided in the following, Hp referring to the useful power is denoted with Hp_{use}. The parameter He is the ratio of electric energy consumed over the total energy and it characterizes the so-called energy split. The quantity He is evaluated along a specified segment or mission. The parameter He_{block} refers, for instance, to the block mission. The need for the dual set of parametric descriptors is elucidated by Isikveren et al. [63] considering the following examples:

- Conventional kerosene based gas turbine propulsion system are described by a Hp and a He equal to zero.
- Pure serial hybrid-electric architecture where only electrical power is provided at the propulsive devices but the energy storage is solely kerosene based. In this case Hp equals 1 and He equals 0.
- Universally-electric aircraft where the energy storage is batteries only which is characterized by an Hp of 1 and an He of 1.

In addition to the algebraic descriptors characterizing the type of propulsion system, the establishment of figure-of-merits for the assessment of the vehicular efficiency is essential. Related to their instantaneous form, they are used for flight technique optimization to determine optimum altitude-speed technique as a function of the aircraft gross-weight, the aerodynamic efficiency and the overall propulsion system efficiency. The integrated form of the metrics along a given mission enables comparing the efficiency of different vehicles to complete an identical transport task. The traditional figure-of-merit used for vehicular efficiency assessment of fuel-based aircraft is the Specific Air Range (SAR). It characterizes the distance traveled per unit of fuel consumed. Optimizing an aircraft for maximum SAR results in minimizing its fuel consumption. This metric is however limited to aircraft using an energy type characterized by a mass flow. A generalization of the SAR was introduced by Seitz et al. [65] with the Energy Specific Air Range (ESAR) which determines the distance traveled per energy consumed. Maximizing ESAR results in minimizing the energy consumption of the aircraft. Optimizing for instance a universally-electric aircraft with respect to ESAR results in minimizing its electrical energy consumption. To enable the optimization of hybrid-energy transport aircraft for minimum energy cost, the COst Specific

Air Range (COSAR) was published by Pornet et al. [66]. The cost of the energy is not the only factor contributing to the total operating cost of an aircraft. Fixed costs and time dependent costs need to be also taken into account. Interested in minimizing the overall cost, airlines base their aircraft fleet operation on so-called Cost-Index which relates basically the cost of time to the cost of energy. A review of the Cost-Index traditionally used for fuel-based aircraft and the establishment of Cost-Index metric for hybrid-energy aircraft are found in [66].

4.2. Aircraft retrofit with hybrid-electric propulsion system

Retrofitting an existing aircraft with a hybrid-electric propulsion system enables having first insights about design parameters and constraints, the behavior of the propulsion system as well as in establishing the right interfaces between the propulsion system and any other aircraft systems. An under-wing podded twin engine narrow-body transport aircraft retrofitted with a hybrid-electric propulsion was investigated by Pornet et al. [21]. The retroïňĄtting of the propulsion system consists of mounting an electric motor powered bybatteries in parallel to the low-pressure shaft of the gas-turbine to supportits operation during some segments of the mission. Due to the intrinsic nature of an aircraft retrofit, the certified Maximum Take-Off Weight (MTOW) of the baseline aircraft represents a design constraint not to be exceeded. The electrical system components including the batteries represent additional weight items. To provide the installation of the electrical system at aircraft level and still respect the MTOW limitation, the hybrid-electric propulsion system is utilized in off-design operations. As the baseline aircraft takes-off at a Take-Off Weight (TOW) lower than MTOW since less fuel is required to perform the off-design stage length, the delta weight available between TOW and MTOW makes the installation the electrical system possible. Sizing of the hybrid-electric propulsion system is the result of an interplay between the maximum power of the installed electric motor, which mainly determines the mass of the electrical system, the total battery mass required, the potential fuel burnreduction and the battery State-Of-Charge (SOC). The SOC characterizes the amount of energy available relative to the total energy of the battery. In order to protect the battery from any damage and to prolong design service goal suitable for use in aerospace, the battery must not be discharged below a certain SOC limit typically set at 20%. The design parameters and constraints are illustrated in the design chart in Figure 4.

It represents the sizing characteristics of the hybrid-electric propulsion system for the case of utilizing the electric motor during cruise only assuming a battery speciïňĄc energy of 1500 Wh/kg at cell-level. The objective of this assessment was to achieve minimum fuel consumption at a stage length corresponding to peak in the utilization spectrum. For the baseline aircraft, the maximum utilization stage length was found to be 900 nm (1667 km). The total installed maximum power of the electric motor $P_{maxEM,totalinst.}$ is varied between 4 to 8 MW and the power setting of the electric motor during cruise $(P/P_{max})_{EM,cruise}$ between 0% to 100%. To maximize the number of installed batteries, the TOW of the hybrid-electric aircraft was set equal to MTOW. The total mass of the installed battery was consequently an outcome of the sizing process. When the electric motor is not used during cruise as indicated by the power setting of 0%, the hybrid-electric aircraft consumes more fuel than the baseline aircraft due to the higher aircraft weight. When increasing the power setting of theelectric motor, more electric energy is consumed increasing the potential block fuel reduction. According to the electric maximum power installed, the battery SOC can become

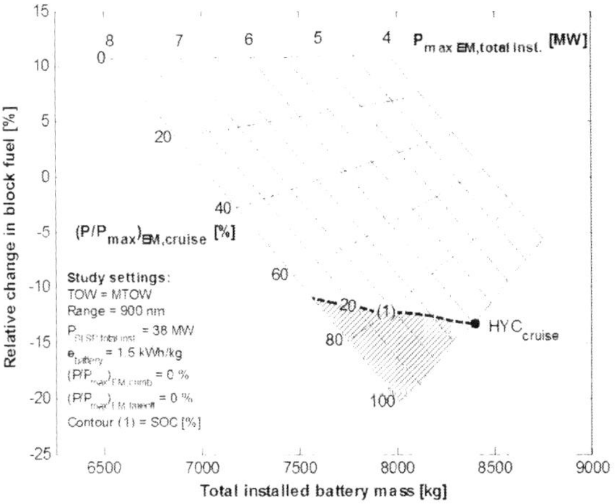

Figure 4. Design chart for hybrid-electric propulsion system in cruise [21]

a limiting factor. The optimum sizing of the hybrid-electric propulsion system in view of achieving minimum fuel burn, indicated in the chart by HYC_{cruise}, results in selecting the maximum power of the electric motor which can be used at 100% under the constraint of the 20% SOC limit of the battery. In the context of this analysis, the optimum sizing results in a 13% block fuel reduction compared to the baseline aircraft by installing a total electric motor power of 6 MW and 8400 kg of batteries. Further investigation presented in [21] demonstrated a potential block fuel reduction of 16% when utilizing the electric motor during climb and cruise. The prospects in block fuel reduction depend strongly on the assumption made in terms of battery technology level. For a battery speciïňAc energy of 1000 Wh/kg at cell-level,results shown that the potential reduction block fuel were reduced by almost 50% compared to results obtained for 1500 Wh/kg [21].

In view of examining the sizing effects resulting from the integration of hybrid-electric propulsion at aircraft level and of investigating the potential market range application of hybrid-electric aircraft, clean sheet design of hybrid-electric aircraft need to be considered.

4.3. Hybrid-electric clean sheet design

A hybrid-electric clean sheet design was investigated on a twin engine narrow body transport aircraft by Pornet et al. [22]. The topological approach of the hybrid-electric propulsion system is identical to the one presented in Section 4.2. Driving an electric motor on the low-pressure shaft has strong influences on the gas-turbine operational characteristics as highlighted in Section 2.1.2. These aspects become more predominant with increasing electric motor power. In order to not modify the contemporary design heuristics of the gas-turbine and to avoid the negative operational influence of the electric motor, the operational strategy selected in this design was to switch-off the gas-turbine during cruise while the electric motor drives by itself the shaft of the propulsor. By equipping only one gas-turbine with an electric motor, the useful degree-of-hybridization Hp_{use} achieved during cruise is 50%.

The other segments of the mission are performed conventionally using the two gas-turbines. By comparing the integrating performance of the hybrid-electric transport aircraft sized for interval design ranges between 500 nm (926 km) and 2100 nm (3889 km) against a suitably projected conventional twin-engine narrow-body aircraft, the prospects in terms of block fuel reduction, change in aircraft size and in vehicular efficiency were investigated for different range applications. In addition, the effect of the battery technology level was assessed for gravimetric specific energy at cell-level varying between 750 Wh/kg and 1500 Wh/kg.

The analysis of the relative change in block fuel versus the relative change in block ESAR is illustrated in Figure 5. When assuming a battery gravimetric specific energy of 1500 Wh/kg, the highest potential block fuel burn reduction of 20% was achieved at a design range of 1100 nm (2037 km). Block fuel reduction is achieved due to the utilization of the electrical energy and because of the increase in overall propulsion system efficiency resulting from the use of the highly efficient electrical system. An improvement in overall propulsion system efficiency of 30% during cruise was evaluated. At 1100 nm (2037 km), around a neutral change in vehicular efficiency was reached compared to an advanced gas-turbine only reference aircraft. It was observed that for increasing design ranges the potential inblock fuel reduction decreases, however, the vehicular efiňĄciency is signiiňĄcantly diminished. In other words, the hybrid-electric transport aircraft requires more energy than the reference aircraft for the same transport task. This trend is explained by the increasing electrical energy demand during cruise, which results in increasing the total battery mass required. The resulting sizing cascade effects, leading to a large increase in MTOW as indicated in Figure 6, are the main cause of the block ESAR reduction. At 1100 nm (2037 km), the MTOW of the hybrid-electric aircraft is increased by 25% compared to an advanced gas-turbine only aircraft. While remaining energy neutral, the block fuel potential reduces for lower design ranges. It is explained by the fact that less electrical energy is used during the reduced portion of the cruise and consequently less block fuel reduction can be achieved. Because of the less pronounced sizing effects and the improvement in overall propulsive efficiency, the total energy consumption remains about neutral. When evaluating hybrid-electric aircraft, it is important to focus not only on fuel burn reduction but to consider also the overall energy consumption. The consumption of electrical energy will affect the operating cost of the hybrid aircraft with respect to the fluctuation of the electric energy price and moreover as the electrical energy might certainly not be produced only through renewable energy sources, its production will impact any carbon life cycle assessments that are undertaken.

Identical trends were observed for lower battery technology levels. However, a lower battery gravimetric specific energy results, for the same amount of energy required, in higher battery mass requirement, which considerably amplifies the sizing cascading effects. It leads consequently in a diminution of the potential fuel burn reduction and in a stronger degradation of the vehicular efficiency. Moreover, the design range at which the largest fuel consumption reduction occurs is reduced to 750 nm (1389 km) and to 900 nm (1667 km) at a specific energy of 750 Wh/kg and 1000 Wh/kg respectively. At these points the block fuel reduction is 9% and 14% and the change in block ESAR is -7% and -4% respectively. In these regards, the potential segment application for this concept turns out to be the regional market. In the context of this investigated concept, a battery technology level of at least 1000 Wh/kg should be reached to enable significant emissions reduction.

The integration of annexed technologies including for instance aerodynamic tailoring technology and flexible, adaptive structures [3] and the consideration of novel aircraft morphologies could lead to improvement in vehicular efficiency. However, the purpose of

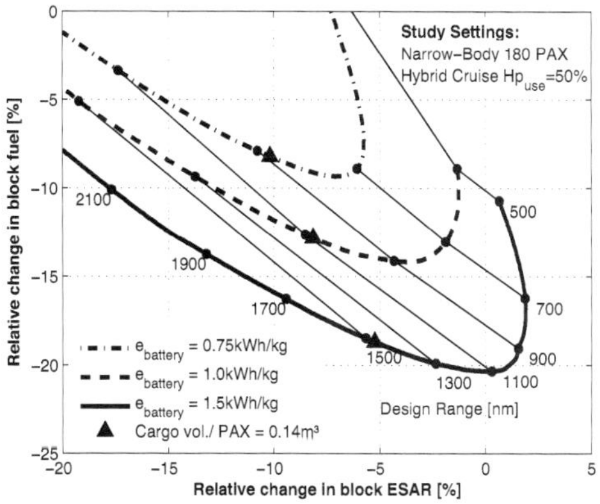

Figure 5. Relative change in block fuel and in block ESAR for a Hp_{use} of 50% in cruise [22]

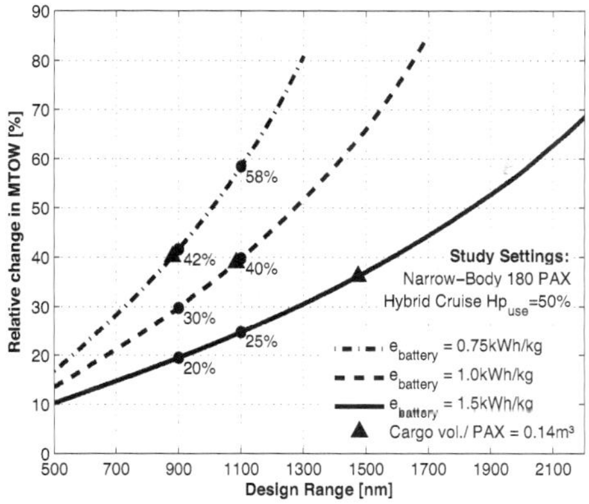

Figure 6. Relative change in MTOW for a Hp_{use} of 50% in cruise [22]

this study was to capture a true potential resulting from the integration of hybrid-electric propulsion system compared to gas-turbine only aircraft.

Another aspect investigated in this study is the volumetric constraint for the housing of the battery within the cargo of the fuselage. According to an assumed density of 1000 kg/m^3 including the volume of the battery, of the thermal management and of the housing, the provision of a standard cargo volume per PAX for regional aircraft of 0.14 m^3 is indicated by a triangle in the Figures 5 and 6. For instance, assuming a battery specific energy of 1000 Wh/kg, the concept is volumetrically constrained for design ranges above 1100 nm (2037 km). Possible evolution of the fuselage geometry, with minor-to-modest aerodynamic and mass penalties, towards double-bubble cross-section could be conceived to free up the design space from this volumetric limitation.

The analysis of this clean sheet design reveals shifts in aircraft design paradigm due to the nature of the hybrid-electric propulsion system as fuel burn reduction can be achieved while the MTOW of the aircraft is increased and moreover fuel burn reduction does not mean automatically an improvement in vehicular efficiency.

While this investigation was aimed to get first insights into the sizing impact of hybrid-electric propulsion system at aircraft level, the full benefit of hybrid-electric technology will be reached through holistic integration of the propulsion system at aircraft level. As defended by Moore and Fredericks [54], the full potential of hybrid-electric aircraft will be demonstrated only once the synergistic benefits from the integration of the hybrid-electric propulsion system at aircraft level are fully understood. In this regard, an innovative hybrid-electric approach is proposed in the next section with the assessment of a distributed hybrid-electric clean sheet design.

4.4. Distributed hybrid-electric clean sheet design

Motivated by the search for higher synergies in the integration of the hybrid-electric propulsion system at aircraft level and by the interest of investigating the influence of increasing Hp on overall aircraft sizing, integrated performance and flight technique optimality, a hybrid-electric narrow-body transport aircraft employing a quad-fan arrangement was investigated by Pornet and Isikveren [23]. Featuring two conventional Geared-TurboFans (GTF) and two Electric Fans (EF), this rudimentary form of distributed propulsion offers numerous advantages compared to previously investigated hybrid-electric architecture and potentials for further evolution as enumerated in Section 2.1. Due to the greater mean time between failure of electric motor compared to gas-turbine, the electric fans are positioned on the outboard to reduce the one engine inoperative implications on performance and sizing. The sizing strategy of the hybrid-electric propulsion system was analyzed as a function of Hp_{use} [23]. The operational strategy selected was to operate the EFs at maximum thrust during the mission segments. Operational phases covering taxi-in/out, descent, landing and hold are performed only with the GTFs. The GTFs are throttled back during cruise to adjust to the instantaneous thrust requirement. This operation was assessed to be suitable up to an Hp_{use} of 45%. Above this value, the efficiency of the gas-turbine in cruise is impaired because of part-load operation resulting from the GTF thrust throttling. By sizing the aircraft for interval design ranges between 900 nm (1667 km) and 2100 nm (3889 km) and for increasing Hp_{use}, the prospects were investigated in terms of potential fuel burn reduction (Figure 7), change in vehicular efficiency (Figure 9) and change in aircraft size (Figure 11). The integrated performance are contrasted against an advanced twin-engine transport aircraft. As indicated in Figure 7, by increasing Hp_{use}, a large reduction in block fuel can be achieved due to the greater utilization of electrical energy and the improvement in overall propulsion system efficiency. Due to the utilization of batteries and because of the electrical system weight, growing Hp_{use} leads to large increase in aircraft weight as illustrated in Figure 11. This effect comes at the detriment of the vehicular efficiency which reduces with increasing Hp_{use} as indicated in Figure 9. The amplification of the sizing cascading effects, linked to the higher electrical energy requirement, explains the degradation of the integrated performance at higher design ranges. From this analysis it can be concluded that the regional market segment is the most suited segment for the application of this hybrid-electric quad-fan transport aircraft with design range between 900 nm (1667 km) to 1300 nm (2408 km). The volumetric constraint for the housing of the investigated battery

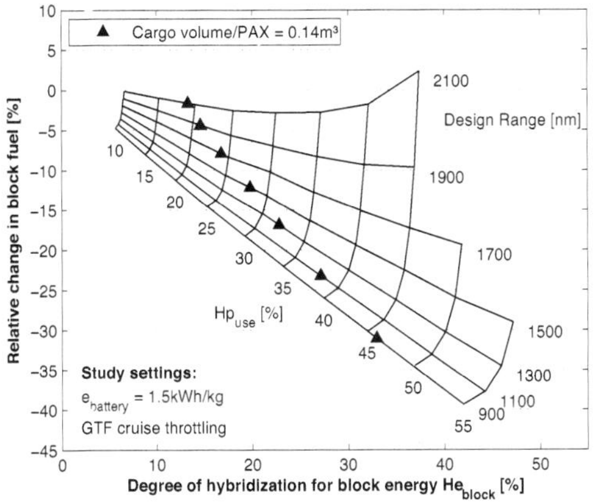

Figure 7. Relative change in block fuel versus block degree-of-hybridization for energy He_{block}. Geared-turbofan cruise throttling [23]

indicates that the cross-section of the narrow-fuselage might limit the design space. With respect to a cargo volumetric constraint of 0.14 m^3 per PAX, a standard volume allocation for regional aircraft, it was found that assuming a gravimetric specific energy of 1500 Wh/kg at cell-level, a block fuel reduction of 15% could be achieved at a design range of 1300 nm (2408 km) and an Hp_{use} of 30% while the vehicular efficiency is degraded by 6%.

To gain insights into the sensitivity with regards to battery technology, integrated performance was investigated for a battery specific energy of 1000 Wh/kg [23]. As indicated in Section 4.3, sizing effects are considerably amplified for lower battery specific energies due to the higher battery mass required for a given energy demand. It results in a diminution of the potential block fuel reduction and a stronger degradation of the vehicular efficiency. The increase in energy demand with growing design range leads to more pronounced degradation of the integrated performance at greater stage lengths. Assuming a gravimetric specific energy of 1000 Wh/kg, no significant block fuel reduction was achieved for design range above 1300 nm (2408 km), whereas, correspondingly for a specific energy of 1500 Wh/kg it is above a design range of 1900 nm (3519 km).

In the outlook proposed in [23], the analysis of a different operational strategy of the hybrid-electric propulsion system was highlighted. The strategy which consists of throttling the EFs during cruise while the GTFs are operated closed to their peak efficiency is investigated in this section. The comparison of the implication of the different strategies is based on the degree-of-hybridization for block energy He_{block}. As this integrated metric includes the overall efficiency chain of the propulsion system and characterizes the integrated block value of the energy split, it is of particular relevance in order to compare the two different operational strategies during cruise. The integrated performance is compared to the identical advanced twin-engine narrow-body aircraft. The potential block fuel reduction, the change in vehicular efficiency and the change in aircraft weight are indicated in Figures 8, 10 and 12 respectively.

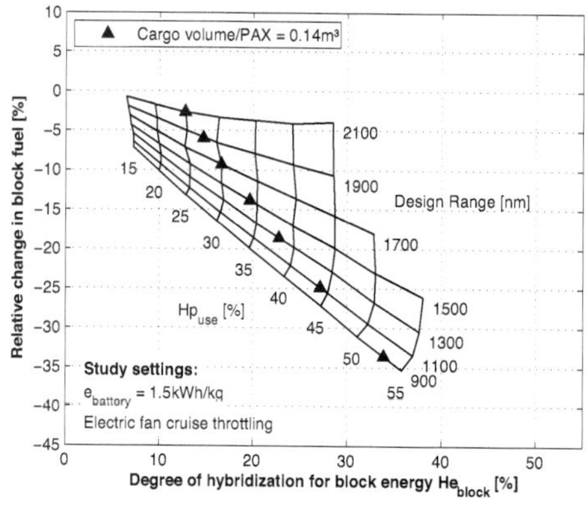

Figure 8. Relative change in block fuel versus block degree-of-hybridization for energy He_{block}. Electric fan cruise throttling

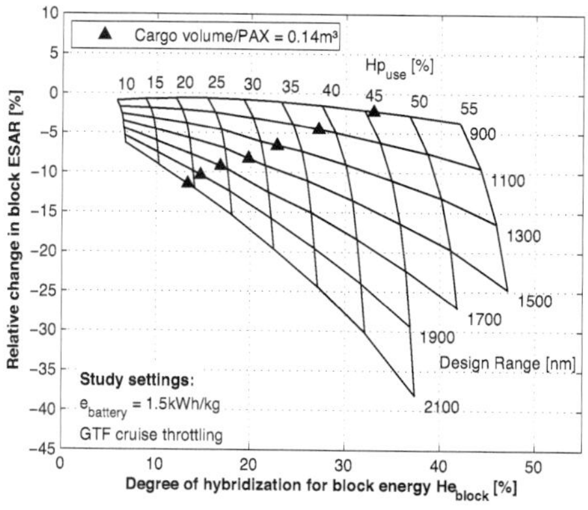

Figure 9. Relative change in block ESAR versus block degree-of-hybridization for energy He_{block}. Geared-turbofan cruise throttling [23]

Interestingly, for a given He_{block} the benefit in terms of block fuel reduction achieved are about identical for both strategies (see Figure 7 and Figure 8). However, it is highlighted that for an identical He_{block} the level of Hp_{use} is larger in the case of throttling the EF during cruise. It means basically that for achieving the same block energy split, a larger electric motor power needs to be installed. This trend is understandable as less electrical energy is consumed during cruise when the EF is throttled back, compared to the first strategy for an identical Hp_{use}. In order to achieve the same He_{block}, a larger electric motor needs consequently to be installed to achieve the same block energy split for an identical block mission. This is the reason which explains the more "compact" carpet plots obtained when selecting the strategy of throttling the EF during cruise. Indeed, for the same variation of Hp_{use} less electrical energy is utilized during cruise resulting in lower value of He_{block}.

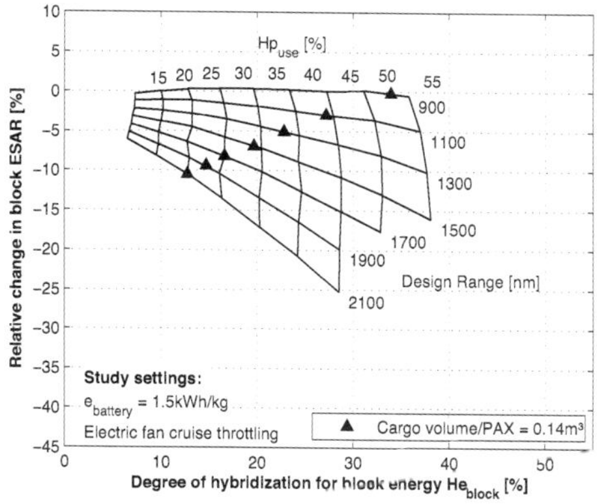

Figure 10. Relative change in block ESAR versus block degree-of-hybridization for energy He_{block}. Electric fan cruise throttling

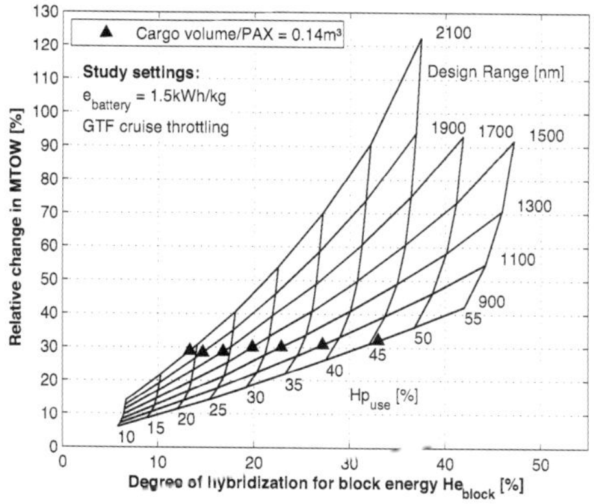

Figure 11. Relative change in MTOW versus block degree-of-hybridization for energy He_{block}. Geared-turbofan cruise throttling [23]

The analysis of the relative change in block ESAR versus He_{block} in Figures 9 and 10 reveals interesting trends resulting from the system implications of the different strategies. As highlighted in the beginning of this section considering the first strategy, an increase in Hp_{use} results in a higher level of thrust throttling of the GTFs during cruise due to the installation of larger EF. This effect contributes to a reduction of the GTF efficiency due to stronger part load operations. This operational consequence can be observed through the noticeable degradation in ESAR at higher levels of Hp_{use} indicated in Figure 7. In the second strategy, the EFs are throttled back and the GTFs run close to their maximum efficiency during cruise. As a result, block ESAR increases slightly with Hp_{use} as the overall propulsion system efficiency is improved through the use of the efficient electrical system and it remains almost independent of Hp_{use} for short design ranges. The decrease in block ESAR at higher

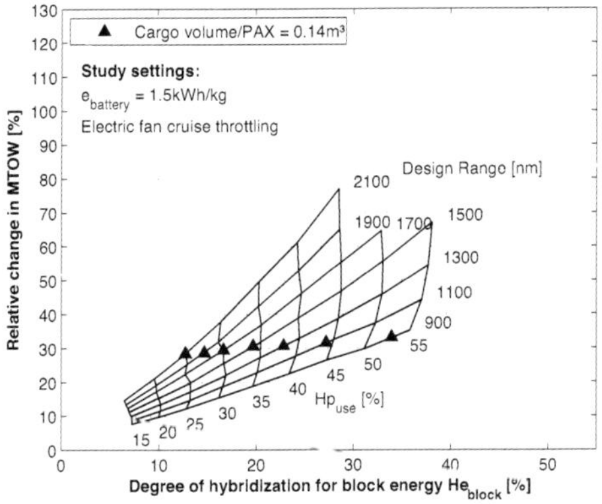

Figure 12. Relative change in MTOW versus block degree-of-hybridization for energy He_{block}. Electric fan cruise throttling

design ranges with increasing levels of Hp_{use} is attributable to sizing cascade effects resulting from the higher electric energy requirement which leads to large increase in aircraft mass (see Figure 12). However, it must be noted that for a given He_{block} the difference in delta block ESAR is small when comparing the different strategies. It is important to highlight at this point that in the current implemented electrical system model, the efficiency of the electrical components, with the exception of the battery, is assumed invariant with respect to the operational conditions and operating time. This assumption is made under the premise of an appropriate thermal management of the electric components and a thoughtful layout of the propulsion architecture. The efficiency of the electrical propulsion system chain in the model depends consequently only on the variation of the battery efficiency with respect to its discharge characteristic and upon the ducted-fan efficiency according to the flight state and the power setting. Moreover, in the current model the speciïňAc weight of the electrical components wereconsidered independent of any scale effect. With the availability of more detailed electrical system models, the dependance of the electrical components efficiency with respect to the altitude-temperature envelope and power load conditions as well as possible variations of the specific weight with scaling effects would be considered.

The impact of the hybrid-electric propulsion on aircraft size according to the EF cruise throttling strategy is illustrated by the change in MTOW versus the change in He_{block} in Figure 12. Similar trends in MTOW change between the different strategies with respect to increase in design range and growing Hp_{use} were identified. For an identical He_{block}, similar values in relative change in MTOW were observed.

In summary, a similar level of reduction in block fuel can be achieved when selecting the throttling of the EFs during cruise. This second operational strategy results in similar change in block fuel and block ESAR as well as in MTOW for an identical He_{block}. However, to reach the same potential in block fuel reduction, a higher level of Hp_{use} (in other words a higher useful electric power relative to the total useful power) needs to be achieved. This translates into the installation of a larger electric motor power. This system implication is rooted in

http://dx.doi.org/10.5772/61506

the nature of the operational strategy. When throttling the EFs during cruise less electrical energy is required at an identical level of Hp_{use}.

5. Conclusion and outlook

The aeronautical community is about to revolutionize aircraft propulsion and aircraft design with the introduction of innovative electric drive approaches to the propulsion system. The full benefit of the hybrid and universally-electric aircraft can only be demonstrated by deploying a truly holistic approach to integrate hybrid-electric and universally-electric propulsion systems at aircraft level. Innovative approaches to electrically driven propulsion systems need to be further analyzed and thoughtfully integrated at aircraft level to determine the full potential. The search for synergistic integration of the propulsion system at aircraft level through tightly interlaced coupling with the airframe as well as with the other systems on-board the aircraft are expected to lead to dramatic shifts in contemporary aircraft design paradigms. Complementing the pure technical assessment of the implications of electric drives to transport aircraft, rigorous analyzes need to be expanded to measure the economical merits. The representation of metrics based on economics into the design charts would enable to select best and balance aircraft concept. Beyond the investigation at aircraft system, life cycle assessment needs to be conducted in order to measure the overall impact of electric drives to the complete chain. While for contemporary propulsion systems the conceptual design analysis of the engine can be done almost independently from the airframe, the synergistic analysis and integration of hybrid-electric or universally-electric propulsion system will result from the interdisciplinary work between the propulsion experts, the electrical systems specialists and the aircraft designer. New professional field of specialization such as electro-aero-propulsion experts could emerge from this multidisciplinarity. These interactions will influence not only the internal organization of aircraft conceptual teams but even beyond will redefine the industrial landscape of partnerships and cooperations due to the essential closer involvement of the suppliers in the early development phase of new aircraft product.

Author details

Clément Pornet*

*Address all coresspodence to: clement.pornet@bauhaus-luftfahrt.net

Bauhaus Luftfahrt, Munich, Germany

References

[1] Advisory Council for Aviation Research and Innovation in Europe (ACARE). Strategic Research and Innovation Agenda (SRIA) - Volume 1. Technical report, Brussels, 2012.

[2] NASA Environmentally Responsible Aviation. N+2 Advanced Vehicle Concepts NRA Draft Solicitation Synopsis. Technical report, 2012.

[3] Askin T. Isikveren and Michael Schmidt. Future Transport Aircraft Ultra-Low Emissions Technology Options. In *GARS Workshop Air Transport and Climate Change*, Worms, Germany, 2014.

[4] Askin T. Isikveren, Arne Seitz, Patrick C Vratny, Clément Pornet, Kay O Plötner, and Mirko Hornung. Conceptual Studies of Universally-Electric Systems Architectures Suitable for Transport Aircraft. In *Deutscher Luft- und Raumfahrt Kongress, DLRK*, Berlin, 2012. Bauhaus Luftfahrt.

[5] Clément Pornet. Hybrid and Universally-Electric Aircraft Concepts. *AccessScience, McGraw-Hill Yearbook of Sciences and Technology (McGraw-Hill Education, 2014), http://dx.doi.org/10.1036/1097-8542.YB150553.*

[6] Timothy P Dever, Kirsten P Duffy, Andrew J Provenza, Patricia L Loyselle, Benjamin B Choi, Carlos R Morrison, and Angela M Lowe. Assessment of Technologies for Noncryogenic Hybrid Electric Propulsion. Technical Report January, NASA/Technical Report 2015-216588, Cleveland, Ohio, 2015.

[7] Jeffrey Freeman, Philip Osterkamp, Michael Green, Andrew Gibson, and Benjamin Schiltgen. Challenges and opportunities for electric aircraft thermal management. *Aircraft Engineering and Aerospace Technology*, 86(6):519–524, September 2014.

[8] Susan Liscouët-Hanke. *A Model-Based Methodology for Integrated Preliminary Sizing and Analysis of Aircraft Power System Architectures*. PhD thesis, Institut National des Sciences Appliquée - Université Paul Sabatier, 2008.

[9] James Felder, Hyun Kim, and Gerald Brown. Turboelectric Distributed Propulsion Engine Cycle Analysis for Hybrid-Wing-Body Aircraft. In *47th AIAA Aerospace Sciences Meeting including The New Horizons Forum and Aerospace Exposition*, pages 1–25, Orlando, Florida, January 2009. American Institute of Aeronautics and Astronautics.

[10] Andrew R Gibson, David Hall, Mark Waters, Benjamin Schiltgen, Trevor Foster, Jonathan Keith, and Philippe J Masson. The Potential and Challenge of TurboElectric Propulsion for Subsonic Transport Aircraft. In *48th AIAA Aerospace Sciences Meeting Including the New Horizons Forum and Aerospace Exposition*, number January, Orlando, Florida, 2010.

[11] Gerald V Brown. Weights and Efficiencies of Electric Components of a Turboelectric Aircraft Propulsion System. In *49th AIAA Aerospace Sciences Meeting including the New Horizons Forum and Aerospace Exposition*, Orlando, Florida, 2011.

[12] Hyun Dae Kim, James L. Felder, Michael T. Tong, Jeffrey J. Berton, and William J. Haller. Turboelectric distributed propulsion benefits on the N3-X vehicle. *Aircraft Engineering and Aerospace Technology*, 86(6):558–561, September 2014.

[13] Devaiah Nalianda and Riti Singh. Turbo-electric distributed propulsion: opportunities, benefits and challenges. *Aircraft Engineering and Aerospace Technology*, 86(6):543–549, September 2014.

[14] Holger Kuhn, Arne Seitz, Larissa Lorenz, Askin T. Isikveren, and Andreas Sizmann. Progress and Perspectives of Electric Air Transport. In *28th International Congress Of The Aeronautical Sciences, ICAS*, Brisbane, 2012.

[15] Christopher A Snyder, Jeffrey J Berton, Gerald V Brown, James L Dolce, Narayan V Dravid, Dennis J Eichenberg, Joshua E Freeh, Christopher A Gallo, Scott M Jones, Krishna P Kundu, Cecil J Marek, Marc G Millis, Pappu L Murthy, Timothy M Roach, Timothy D Smith, George L Stefko, Roy M Sullivan, Robert T Tornabene, Karl A Geiselhart, and Albert Kascak. Propulsion Investigation for Zero and Near-Zero Emissions Aircraft. Technical Report May, NASA/TM-2009-215487, National Aeronautics and Space Administration, Cleaveland, Ohio, 2009.

[16] Patrick Christoph Vratny, Corin Gologan, Clément Pornet, Askin T. Isikveren, and Mirko Hornung. Battery Pack Modeling Methods for Universally-Electric Aircraft. In *4th CEAS Air & Space Conference*, pages 525–535, Linköping, Sweden, 2013. Linköping University Electronic Press.

[17] Holger Kuhn, Christoph Falter, and Andreas Sizmann. Renewable Energy Perspectives for Aviation. In *Proceedings of the 3rd CEAS Air&Space Conference and 21st AIDAA Congress*, volume 1259, pages 1249–1259, Venice, Italy, 2011.

[18] Airbus Group. The future of e-aircraft, accessed april 2015, http://www.airbusgroup.com.

[19] Marty K. Bradley and Christopher K Droney. Subsonic Ultra Green Aircraft Research : Phase I Final Report. Technical Report April, Huntington Beach, California, 2011.

[20] Marty K. Bradley and Christopher K. Droney. Subsonic Ultra Green Aircraft Research Phase II : N + 4 Advanced Concept Development. Technical report, NASA, Huntington Beach, California, 2012.

[21] Clément Pornet, Corin Gologan, Patrick C. Vratny, Arne Seitz, Oliver Schmitz, Askin T Isikveren, and Mirko Hornung. Methodology for Sizing and Performance Assessment of Hybrid Energy Aircraft. *Journal of Aircraft, Vol. 52, No. 1 (2015), pp. 341-352. doi: 10.2514/1.C032716*, pages 1–12, 2014.

[22] Clément Pornet, Sascha Kaiser, Askin T. Isikveren, and Mirko Hornung. Integrated Fuel-Battery Hybrid for a Narrow-Body Sized Transport Aircraft. *DOI 10.1108/AEAT-05-2014-0062, Aircraft Engineering and Aerospace Technology, Vol. 86, Iss. 6, pp. 568-574*, 86(6), 2014.

[23] Clément Pornet and Askin T. Isikveren. Conceptual Design of Hybrid-Electric Transport Aircraft. *Journal of Progress in Aerospace Sciences* 2015, DOI: 10.106/j.paerosci.2015.09.002

[24] R. W. Schurhoff. *The Development and Evaluation of an Optimal Powertrain Control Strategy for a Hybrid Electric Vehicle*. PhD thesis, University of California - Davis, CA, 2002.

[25] Frederick G. Harmon, Andrew A. Frank, and Jean-Jacques Chattot. Conceptual Design and Simulation of a Small Hybrid-Electric Unmanned Aerial Vehicle. *Journal of Aircraft*, 43(5):1490–1498, September 2006.

[26] Oliver Schmitz and Mirko Hornung. Methods for Simulation and Analysis of Hybrid Energy Propulsion Systems. In *Proceedings of the 62nd Deutscher Luft- und Raumfahrt Kongress (DLRK)*, Stuttgart, 2013.

[27] Oliver Schmitz. Fahrzeugtriebwerk, Fahrzeug mit diesem Fahzeugtriebwerk und Verfahren zum Betrieb dieses Fahrzeugtriebwerkes, 2012.

[28] Oliver Schmitz and Mirko Hornung. Unified Applicable Propulsion System Performance Metrics. *J. Eng. Gas Turbines Power*, 135(11):1–9, 2013.

[29] K.C. Divya and Jacob Ostergaard. Battery energy storage technology for power systems: an overview. *Electric Power Systems Research*, 79:511–520, April 2009.

[30] Bruno Scrosati and Jürgen Garche. Lithium batteries: Status, prospects and future. *Journal of Power Sources*, 195:2419–2430, May 2010.

[31] B. Ketterer, U. Karl, D. Möst, and S. Ulrich. Lithium-Ion Batteries: State of the Art and Application Potential in Hybrid-, Plug-In Hybrid- and Electric Vehicles. Technical report, Forschungszentrum Karlsruhe in der Helmholtz-Gemeinschaft, Institut für Materialforschung I, 2009.

[32] William H. Woodford, R. Alan Ransil, and Yet-Ming Chiang. Advanced Batteries: Beyond Li-ion. Technical report, NPC Future Transportation Fuels Study, 2012.

[33] Jake Christensen, Paul Albertus, Roel S. Sanchez-Carrera, Timm Lohmann, Boris Kozinsky, Ralf Liedtke, Jasim Ahmed, and Aleksandar Kojic. A Critical Review of Li-Air Batteries. *Journal of The Electrochemical Society*, 159, 2012.

[34] Alexander Kraytsberg and Yair Ein-Eli. Review on LiâĂŞair batteries: Opportunities, limitations and perspective. *Journal of Power Sources*, 196:886–893, February 2011.

[35] S Visco, E Nimon, B Katz, M Chu, and L De Jonghe. High Energy Density Lithium-Air Batteries with No Self Discharge. In *in Proceedings of the 42nd Power Sources Conference*, pages 201–203, 2006.

[36] Lonnie Johnson. The Viability of High Specific Energy Lithium Air Batteries. In *Symposium on Research Opportunities in Electrochemical Energy Storage - Beyond Lithium Ion: Materials Perspective*, Tennessee, USA, 2010.

[37] Axel Thielmann. Technologie-Roadmap Lithium-Ionen-Batterien 2030. Technical report, Fraunhofer-Institut für System- und Innovationsforschung ISI,, Karlsruhe, 2010.

[38] Philippe J Masson, Gerald V Brown, Danielle S Soban, and Cesar A Luongo. HTS machines as enabling technology for all-electric airborne vehicles. *Superconductor Science and Technology*, 20(8):748–756, August 2007.

[39] Cesar A Luongo, Philippe J Masson, Taewoo Nam, Dimitri Mavris, Hyun D Kim, Gerald V Brown, Mark Waters, and David Hall. Next Generation More-Electric Aircraft: A Potential Application for HTS Superconductors. *IEEE Transactions on applied superconductivity*, 19(3):1055–1068, 2009.

[40] Philippe J Masson and Cesar A Luongo. HTS Machines for Applications in All-Electric Aircraft. In *Proceedings of the PES Meeting*, Tampa, Florida, 2007.

[41] K. Sivasubramaniam, T. Zhang, M. Lokhandwalla, E. T. Laskaris, J. W. Bray, B. Gerstler, M. R. Shah, and J P Alexander. Development of a High Speed HTS Generator for Airborne Applications. *IEEE Transactions on applied superconductivity*, 19(3):1656–1661, 2009.

[42] Peter Malkin and Meletios Pagonis. Superconducting electric power systems for hybrid electric aircraft. *Aircraft Engineering and Aerospace Technology*, 86(6):515–518, September 2014.

[43] Paul M. Ordin. Safety Standard for Hydrogen and Hydrogen Systems. Technical report, NASA NSS1740.16, National Aeronautics and Space Administration, 2005.

[44] A Westenberger. Liquid Hydrogen Fuelled Aircraft - System Analysis. Technical Report May 2002, Final Technical Report, Cryoplane Project, 2003.

[45] The Boeing Commercial Airplane Company. An exploratory study to determine the integrated technological air transportation system gound requirements of liquid-hydrogen-fueled subsonic, long-haul civil air transports. Technical Report September 1976, NASA Contractor Report, 1976.

[46] Lockheed Company. LH2 Airport Requirements study. Technical report, NASA Contractor Report, 1976.

[47] Panagiotis Laskaridis. *Performance investigations and systems architectures for the More Electric Aircraft*. PhD thesis, Crandfield University, 2004.

[48] Howard Smith. Airframe integration for an LH2 hybrid-electric propulsion system. *Aircraft Engineering and Aerospace Technology*, 86(6):562–567, September 2014.

[49] Hans-Jörg Steiner, Arne Seitz, Kerstin Wieczorek, Kay Plotner, Askin T Isikveren, and Mirko Hornung. Multi-Disciplinary Design And Feasibility Study of Distributed Propulsion Systems. In *28th International Congress Of The Aeronautical Sciences, ICAS*, Brisbane, 2012.

[50] Askin T Isikveren, Arne Seitz, Julian Bijewitz, Mirko Hornung, Artur Mirzoyan, Alik Isyanov, Jean-luc Godard, Stefan Stückl, and Jan Van Toor. Recent Advances in Airframe-Propulsion Concepts with Distributed Propulsion. In *29th Congress of the International Council of the Aeronautical Sciences*, St. Petersburg, Russia, 2014.

[51] Arne Seitz, Julian Bijewitz, Sascha Kaiser, and Guido Wortmann. Conceptual investigation of a propulsive fuselage aircraft layout. *Aircraft Engineering and Aerospace Technology*, 86(6):464–472, September 2014.

[52] Julian Bijewitz, Arne Seitz, and Mirko Hornung. Multi-disciplinary Design Investigation of Propulsive Fuselage Aircraft Concepts. In *4th EASN Association International Workshop on Flight Physics and Aircraft Design*, Aachen , Germany, 2014.

[53] Askin T Isikveren, Arne Seitz, Julian Bijewitz, Arthur Mirzoyan, Alik Isyanov, Richard Grenon, Olivier Atinault, Jean-luc Godard, and Stefan Stückl. Distributed Propulsion and Ultra-High By-Pass Rotor Study at Aircraft Level. *The Aeronautical Journal (accepted)*, 2015.

[54] Mark D Moore and Bill Fredericks. Misconceptions of Electric Propulsion Aircraft and their Emergent Aviation Markets. In *AIAA doi/0.2514/6.2014-0535, SciTEch*, 2014.

[55] William J Fredericks, Mark D Moore, and Ronald C Busan. Benefits of Hybrid-Electric Propulsion to Achieve 4x Increase in Cruise Efficiency for a VTOL Aircraft. In *AIAA Aviation Technology, Integration, and Operations (ATIO) Conference*, Los Angeles, California, 2013.

[56] William J Fredericks. Impact of Distributed Electric Propulsion (DEP) on Aircraft Design. In *Special Conference on Disruptive Green Propulsion Technologies: Beyond the Competitive Horizon*, number November, London, United Kingdom, 2014.

[57] Alex M Stoll, Joeben Bevirt, Mark D Moore, William J Fredericks, and Nicholas K Borer. Drag Reduction Through Distributed Electric Propulsion. In *Aviation Technology, Integration and Operations Conference*, number June, Atlanta, Georgia, 2014.

[58] Benjamin Schiltgen, Michael Green, and Andrew Gibson. Analysis of Terminal Area Operations and Short Field Performance of Hybrid Electric Distributed Propulsion. In *AIAA Aviation 2013*, volume 86, pages 584–590, Los Angeles, California, September 2013.

[59] Hans-Jörg Steiner, Patrick C Vratny, Corin Gologan, Kerstin Wieczorek, Askin T Isikveren, and Mirko Hornung. Optimum number of engines for transport aircraft employing electrically powered distributed propulsion. *CEAS Aeronautical Journal*, Volume 5(Issue 2):157–170, 2014.

[60] Martin Hepperle. Electric Flight - Potential and Limitations. In *NPU-DLR Workshop*, Braunschweig, 2014.

[61] Stefan Stückl, Jan van Toor, and Hans Lobentanzer. VOLTAIR - The All Electric Propulsion Concept Platform - A Vision For Atmospheric Friendly Flight. In *28th International Congress Of The Aeronautical Sciences, ICAS*, Brisbane, 2012.

[62] J. Paulson T. Banning, G. Bristow, C. Level, L. Sollmann, J. Calderon-Fernandez, D. Wells, M. Olson, N. Davis, C, Du, S. Ambadpudi. 2012-2013 FAA Design Competition for Universities Electric / Hybrid-Electric Aircraft Technology Design Category - NXG-50 - Georgia Tech. Technical report, 2013.

[63] Askin T. Isikveren, Sascha Kaiser, Clément Pornet, and Patrick C. Vratny. Pre-design Strategies and Sizing Techniques for Dual-Energy Aircraft. *DOI: 10.1108/AEAT-08-2014-0122, Aircraft Engineering and Aerospace Technology Journal, Vol. 86, Iss. 6, pp. 525-542*, 2014.

[64] Larissa Lorenz, Arne Seitz, Holger Kuhn, and Andreas Sizmann. Hybrid Power Trains for Future Mobility. In *Paper 1316, Deutscher Luft- und Raumfahrtkongress*, pages 1–17, Stuttgart, Germany, 2013.

[65] Arne Seitz, Oliver Schmitz, Askin T Isikveren, and Mirko Hornung. Electrically Powered Propulsion: Comparison and Contrast to Gas Turbines. In *Deutscher Luft- und Raumfahrt Kongress, DLRK*, Berlin, 2012. Bauhaus Luftfahrt.

[66] Clément Pornet, Sascha Kaiser, and Corin Gologan. Cost-Based Flight Technique Optimization for Hybrid Energy Aircraft. *DOI 10.1108/ AEAT-05-2014-0075, Aircraft Engineering and Aerospace Technology Journal*, 86(6), 2014.

[67] Cronin, M. Advanced Power Generation Systems for More Electric Aircraft. *SAETechnical Paper 912186*, 1991.

[68] Cronin, M. The all-electric aircraft. *IEE Review*, vol. 36, 1990, pp. 309-311.

[69] Quigley, R. E. More Electric Aircraft. *Proceedings of the 8th Applied PowerElectronics Conference and Exposition, APEC 93*, 1993, pp. 906-911.

[70] Faleiro, L. Beyond the More Electric Aircraft. *Aerospace America*, 2005.

[71] Ahmed Abdel-Hafez Power Generation and Distribution System for a More Electric Aircraft - A Review. Recent Advances in Aircraft Technology, Dr. Ramesh Agarwal (Ed.), ISBN: 978-953-51-0150-5, InTech, 2012.

[72] Gohardani, A. S., Doulgeris, G., and Singh, R. Challenges of future aircraft propulsion: A review of distributed propulsion technology and its potential application for the all electric commercial aircraft. *Progress in Aerospace Sciences*, vol. 47, Jul. 2011, pp. 369-391.

[73] Gohardani, A. S. A synergistic glance at the prospects of distributed propulsion technology and the electric aircraft concept for future unmanned air vehicles and commercial or military aviation. *Progress in Aerospace Sciences*, vol. 57, 2013, pp. 25-70.

[74] Gohardani, A. S. (Editor), *Propulsion Technology, ISBN: 978-1-62948-588-1*. Nova Science Publishers, 2014.

[75] Kirner, R., Raffaelli, L., Rolt, A., Laskaridis, P., and Singh, R. Analysis of Distributed Propulsion on Advanced Vision 2020 Aircraft Concepts. *International Society for Air Breathing Engines*, Busan, Korea: 2013.

[76] Mantic-Lugo, V., Doulgeris, G., and Singh, R. Computational Analysis of the Effects of a Boundary Layer Ingesting Propulsion System in Transonic Flow. *Proceedings of the Institution of Mechanical Engineers, Part G: Journal of Aerospace Engineering*, 2012.

Chapter 6

Electric-driven Zonal Hydraulics in Non-Road Mobile Machinery

Tatiana A. Minav, Jani E. Heikkinen and Matti Pietola

Additional information is available at the end of the chapter

http://dx.doi.org/10.5772/61793

Abstract

The goal of this research is to apply direct-driven hydraulics (DDH) to the concept of zonal (i.e., locally and operation-focused) hydraulics, which is an essential step in the hybridization and automation of machines. DDH itself aims to combine the best properties of electric and hydraulic technologies and will lead to increased productivity, minimized energy consumption and higher robust performance in both stationary and mobile machines operating in various environments. In the proposed setup, the speed and position control of a double-acting cylinder is implemented directly with an electric motor drive in a closed-loop system without conventional control valves and an oil tank. The selection of the location of the hydraulic accumulator and connection of the external leakage lines will also be part of this study. Simulations and experimental research to study the details of the hydromechanical and electrical realization of the DDH are performed.

Keywords: direct driven hydraulics, non-road mobile machinery, energy efficiency, electric drive, hybrid topology, energy regeneration

1. Introduction

The climate for industrial and economic growth is changing as resource scarcity pushes up costs, and enterprises are more closely scrutinized because of changing expectations about their public and social responsibilities. Governments have set tight CO_2 emission rules [1] and new exhaust limits have just been implemented, such as Tier Emission Standards. The next known step will be the 2019/2020 Tier V, which is a regulation imposing an ulterior sharp tightening of exhaust limits, especially in terms of particles. Now, there is a four-year time window to prepare engines for the upcoming regulations through solutions for peak power shaving and the downsizing of diesel engines. Electric and hybrid vehicles are a suitable solution for reaching these target environmental requirements; such vehicles have a huge potential for application in the non-road mobile machinery (NRMM) industry, with its market in mining, process and goods manufacturing, forest harvesting and construction work. Figure 1 illustrates typical examples of NRMMs.

Figure 1. Examples of non-road mobile machinery: (a) excavator; (b) mine loader and (c) forest harvester.

The NRMM industry and its customers are traditionally quite conservative in putting new ideas and technologies into use. When they opt for "mild hybridization", the machine manufacturers do not need to change the construction frame of the whole machine immediately. However, some changes in layout design concepts, machine types and business models should be considered in the long term. Therefore, we address several challenges to improve NRMMs by bringing zonal hydraulics into machines. Combining the best properties of traditional hydraulics and electric intelligence allows the following benefits to be achieved in NRMMs:

- easy electrification of NRMMs
- higher efficiency compared to conventional machines
- electrohydraulic power pack with no tank and pipelines
- reduction of potential leakage points
- power-on-demand control
- sensorless position control

1.1. State of the art

During recent years, the NRMM industry has had a tendency towards integrated, compact electrohydraulic systems that deliver powerful, linear movement with either valve- or pump-controlled systems. These technological steps are considered important stages towards a solution to commonly identified problems, such as the reduction of CO_2 emissions and improved performance, productivity, reliability and controllability [2, 3], as well as good options for forcing a smaller diesel engine to operate within its optimum efficiency area since, in general, the diesel engines of NRMMs operate far from their optimum efficiency range [4]. Restrictive environmental regulations are not the only factor imposing the need for innovation in NRMMs. Today, as the search for natural resources expands to areas previously considered too remote and environmentally hostile to support viable extraction and processing operations, researchers all around the world are facing the challenge of establishing and maintaining industrial activities in extreme environments. In these, components of the drive face thermal issues, primarily at start-up, when everything in the drive has cooled down to low, even Arctic temperatures. Under these conditions, conventional hydraulic fluid in all hydraulic systems solidify, so when the NRMM starts up, there is no fluid circulating to protect the pump and other components from extreme wear. For successful operation, a relevant start-up procedure is crucial, as the risk of equipment failure is unacceptable in remote locations.

Zonal or decentralized hydraulics—as we will call it—is an approach first introduced in the aircraft industry [5]. Adapting this time-proven design from the aircraft industry, simplifying its design as well as broadening its orientation and performance options will support a similar development in NRMMs. In a fully zonal system, the hydraulic pumps are removed from the engine and replaced within hydraulic power packs distributed throughout the NRMM system. In this architecture, multiple hydraulic power sources may be utilized in each zone in order to achieve energy savings and work with a power-on-demand approach.

Figure 2 illustrates the zonal hydraulic architecture applied to NRMMs. A challenge in this is the increasing number of electric components in the limited volume available in the vehicle. On the other hand, the main advantages of this architecture are the reduced hydraulic tubing (tubes are replaced with wiring), the elimination of some hydraulic components and simplified machine assembly.

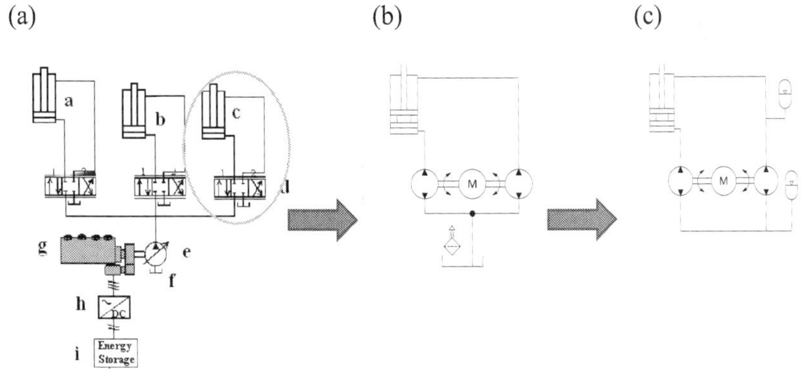

Figure 2. Schematics of (a) conventional NRMM, (b) hybrid NRMM with secondary power source DDH with a conventional tank and (c) DDH without a tank.

In this study, an electrohydraulic actuator (EHA) will be used to achieve high power density, low noise and high performance in a compact package in order to take the development of a power unit onto the next level. The electrohydraulic actuator allows decoupling from the main hydraulic system, thus enabling a zonal approach and, at the same time, reducing parasitic losses in order to obtain better fuel efficiency and lower operating costs. The robust, leak-free, one-piece housing design delivers system simplicity and lowers both installation and maintenance costs. The electrohydraulic actuator reduces space and weight demands. It can eliminate hoses, fittings, valves and fixtures and is easy to integrate into larger systems.

Currently, electrohydraulic actuators are mostly installed and developed for aircraft applications [6, 7], where the price level and reliability requirements are very high. Most of the research studies related to electrohydraulic actuators have been conducted to adjust the state of the servo valve [8, 9]. However, this approach results in low energy efficiency because of the flow via the pressure relief valves of the hydraulic pumps and throttle losses at the control valves. Consequently, several techniques have been developed to overcome this drawback in order to achieve higher efficiency. In ref. [10], the concept of a pump-controlled electrohydraulic actuator is introduced as an advanced hydraulic system where the proposed structure of the electrohydraulic actuator is directly operated by a bidirectional pump. There are already commercially integrated power packages on the market, but they are based on conventional technology and they do not respond to the above-mentioned challenges created by the hybridization problem. Examples of commercial pump-controlled electrohydraulic actuators are, for instance, the mini-motion package from Kayaba Industry Co. [11] and the intelligent hydraulic servo drive-pack from Yuken Kogyo Co. [12]. The disadvantages of the architecture evidenced in ref. [10] are the use of complex and expensive pumps and the lower dynamic properties of the system in general. In addition, none of these commercial solutions are designed for power-on-demand control, sensorless positioning or even energy regeneration, all highly desirable improvements in a competitive product.

Some studies have been conducted to introduce the concept of electrohydraulic actuators as zonal hydraulics in NRMM. In ref. [13], an electrohydraulic actuator is used for the power steering of heavy vehicles. In refs. [14–17], a compact drive for automation of all kinds of linear motions was introduced and investigated from the thermal point of view. In these sources, even if they proposed direct pump control similar to the direct-driven hydraulic approach, a set of valves is used to balance the flow and to ensure the direction of the electrohydraulic actuator motion. In refs. [18, 19], direct-driven hydraulics (DDH), an electrohydraulic actuator, was introduced without conventional directional valves. The DDH drive combines the best properties of electric and hydraulic drive technologies in one:

- direct control of flow, as well as the velocity and position of the actuator
- high hydraulic efficiency because there are virtually no flow-restricting valves
- no need for expensive variable displacement pumps
- easier to find the optimum operational point for each of the powertrain components (fewer compromises) since each actuator has two pump units and an electric motor
- the pump units and internal combustion motor are disconnected
- possibility of realizing efficient energy recovery systems using both hydraulic accumulators and electric batteries

Therefore, the idea of implementing the DDH drive in the powertrain of NRMMs as an application of the zonal hydraulic concept was born. In this research, direct-driven hydraulics

(DDH) is seen as a tool to convert existing NRMMs to hybrids combining the best properties of traditional hydraulics and electric intelligence. Figure 3 illustrates the hybrid proposal for NRMMs. With this approach, several challenges that traditionally manifest themselves in the hybridization process will be solved. In the proposed setup, the speed and position control of a double-acting cylinder is implemented directly with a motor drive in a closed-loop system without conventional control valves and an oil tank. The selection of the optimal size and the location of the hydraulic accumulator will also be part of this study. Simulations and experimental research to study the details of the hydro-mechanical and electrical realization of the DDH will be performed.

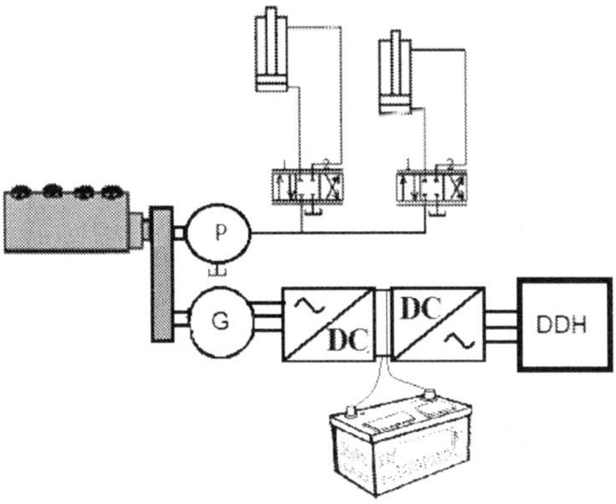

Figure 3. Hybrid proposal for NRMM with DDH.

The next chapter introduces the experimental DDH setup.

2. Description of test setup

The experimental setup uses a speed-controlled electric servo motor drive rotating two hydraulic pumps to directly control the amount of hydraulic oil pumped to the asymmetrical double-acting cylinder MIRO C-10-60/30× 400. The simplified circuit diagram of the experimental test setup is illustrated in Figure 4. The hydraulic pump/motors P_1 and P_2 create an input and output flow that depends on the rotating speed of the servo motor. The oil pressure rises to the required level as determined by the payload. During lowering motion, the potential energy of the payload creates a flow that rotates the hydraulic machine P_1 as a motor and the hydraulic machine P_2 as a pump, and the mechanically connected electric motor acts as a generator, which is controlled by the frequency converter. The speed-controlled generator controls the amount of fluid flow and the position of the payload. The program for the electric drive controls both the electrical and hydraulic sides of the system as there is no conventional valve control.

For easier understanding of the real displacements of the pumps, the following ratios R_A and R_Q will be defined. In this work, the ratio between the cylinder areas is defined by the following equation:

$$R_A = \frac{A_3}{A_1},$$

where $A_3 = A_1 - A_2$ is the cylinder area from the rod side. The diameter of the cylinder piston head is $d_1 = 0.06$ m and $d_2 = 0.03$ m is the diameter of the piston rod, which determines the piston surface areas A_1 and A_2. According to the datasheet, the cylinder stroke is $l = 0.4$ m.

As previously stated, the pump/motors P_1 and P_2 are mounted on the same axis, so their speeds are identical. If the pump leakage is ignored, the ratio of the flow rates R_Q can be produced directly from the displacement of the pump/motors D_1 and D_2.

$$R_Q = \frac{D_2}{D_1}.$$

Figure 4 illustrates the first prototype of the DDH setup with $R_Q \approx 0.63$ and $R_A = 0.75$. To test the concept, the components for this prototype were taken off the shelf.

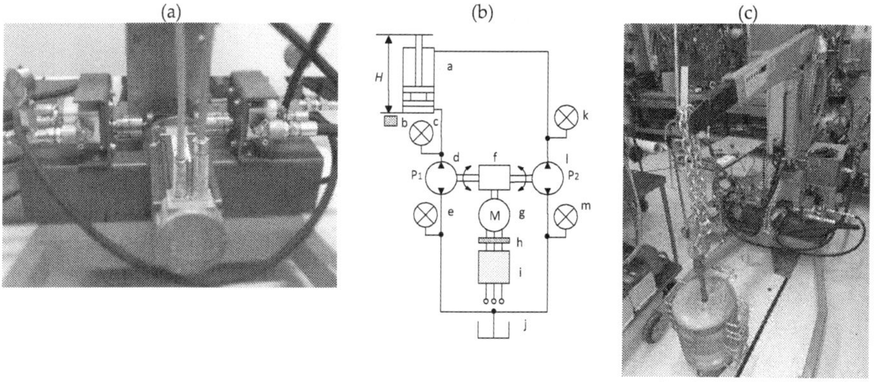

Figure 4. First prototype of the DDH setup: (a) zoom view;(b) the experimental setup consists ofa) a double-acting cylinder, b) wire-actuated encoder, c) pressure sensor, d) reversible gear pump/motor P_1, e) pressure sensor, f) gearbox, g) PMSM motor/generator, h) current sensors, i)frequency converter, j) tank, k) pressure sensor, l) reversible gear pump/motor P_2 and m) pressure sensor in the tank line; (c) the crane is used for the DDH setup as a test platform (side view).

Two XV-2M internal gear pump/motors by Vivoil with displacements of 14.4 and 22.8 cm^3/rev were used, P_2 and P_1, respectively [20]. The position feedback from the motor is given by means of its in-built incremental encoder (4096 pulses per revolution, resolution 14 bits), and read with the Unidrive SP1406 drive software [21]. It converts the AC power supply from the line and allows the speed of the permanent magnet brushless servo motor, Unimotor 115U2C manufactured by Emerson Control Techniques, to be set, taking advantage of the information obtained by the feedback device fitted to ensure the rotor speed is exactly as demanded [22]. This experimental setup was tested with a payload of 150 kg at motor speed ranges from 300 to 500 rpm. Figure 5 shows an example of measured data for a motor speed of 400 rpm and a payload of 150 kg: speed, torque and pressure.

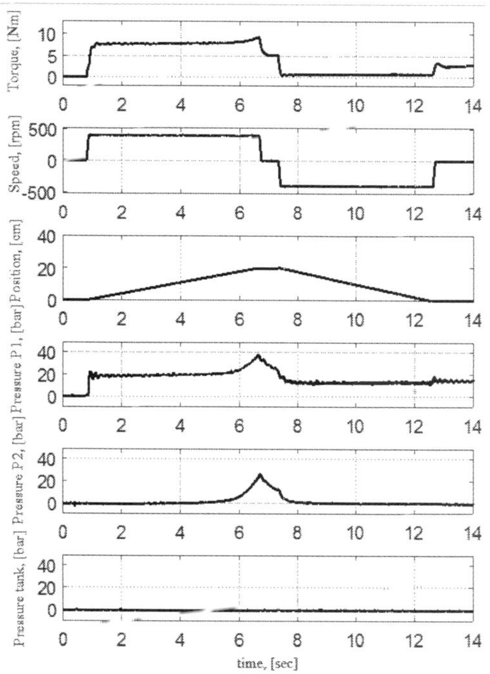

Figure 5. Example of measured data of the first DDH prototype: motor speed, torque, pressure in the pump/motor line P_1 and in the pump/motor line P_2. For a motor speed of 400 rpm and a payload of 150 kg.

In Figure 5, during the lifting, which lasts from 1 to 7 s, the pressure in the pump/motor P_1 is about 2 MPa, which rises by the end to 4 MPa. It is worth remarking that the rise appears to be due to the difference between R_Q and R_A. A similar rise also occurs in the pump/motor line P_2, from atmospheric pressure to 2.6 MPa.

During the lowering, performed from 7 to 13 s, a drop in the pressure from high pressure to about 1 MPa happens. During the lifting, the motor torque is 7 Nm; during the lowering, the torque is around 1 Nm. The pressure in the tank line is close to atmospheric pressure, as it is supposed to be in an open system.

In order to overcome the rise in pressure at the end of the movement and investigate the tankless approach, a new prototype was designed. Figure 6 illustrates the experimental test setup of the second DDH prototype. The system is closed by giving up the tank completely (compare to Figure 2b) and replacing it with a hydraulic accumulator A.

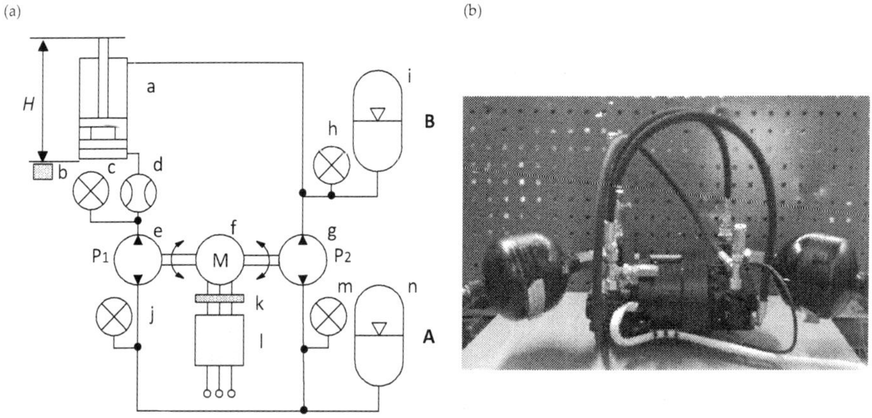

Figure 6. (a) Schematics of the setup of the second DDH prototype: a) double-acting cylinder, b) wire-actuated encoder, c) pressure sensor, d) flow meter,e) reversible gear pump/motor P_1, f) PMSM motor/generator, g) reversible gear pump/motor P_2, h) pressure sensor, i) hydraulic accumulator B, j) pressure sensor in tank line, k) current sensors, l) frequency converter, m) pressure sensor and n) hydraulic accumulator A; (b) photograph of DDH setup.

Within the framework of this work for the second prototype $R_Q \approx 0.73$ and $R_A = 0.75$. It can be seen that the ideal $R_Q = R_A$ was not achieved. Therefore, an accumulator B is added to compensate for the displacement difference between the available pump/motors and cylinder areas.

2.1. Components used

In the second prototype, DDH setup an electric motor, an IndraDyn T MST130A-0250-N torque motor from Bosch Rexroth, is used. The parameters of the electric motor are shown in Table 1.

Rated power	Rated torque	Maximum torque	Nominal speed	Maximum speed	Pole pairs	Torque constant	Voltage constant	Resistance	Inductance
1.2 kW	4.5 Nm	13 Nm	2500 rpm	4000 rpm	10	1.3 Nm/A	0.085 V/ min^{-1}	5.9 Ohm	17.5 mH

Table 1. IndraDyn T MST130A-0250-N parameters [23].

Bosch Rexroth AZMF-12-011U and AZMF-12-008U external gear motors are used as hydraulic pumps/motors. The parameters of the pump/motors are shown in Tables 2 and 3.

Flow rate	Maximum pressure	Leakage-oil pressure	Minimum speed	Maximum speed
11×10^{-6} m³/rev	25 MPa	0.3 MPa*	500 Nm	3500 rpm

Table 2. Parameters of the AZMF-12-011U hydraulic pump/motor P_1 [24].

Flow rate	Maximum pressure	Leakage-oil pressure	Minimum speed	Maximum speed
8×10^{-6} m³/rev	25 MPa	0.3 MPa*	500 Nm	4000 rpm

* Short-term pressure during start 1 MPa.

Table 3. Parameters of the AZMF-12-008U hydraulic pump/motor P_2 [24].

Two Parker AD100B20T9A1 diaphragm accumulators are utilized as the initial choices. Table 4 shows the parameters of the hydraulic accumulators.

Capacity (volume)	Maximum pressure
0.001 m³	200 MPa

Table 4. Parameters of the AD100B20T9A1 hydraulic accumulators A and B.

The pressure was measured with Gems 3100R0400S pressure transducers. The velocity of the cylinder piston was measured with an SGW/SGI wire-actuated encoder by SIKO.

According to the manufacturer, the pump/motors used have an external leakage-oil line with a pressure limitation of 0.3 MPa. During the start, the maximum allowed short-term pressure is 1 MPa. Thus, a detailed investigation of the hydraulic connection of the external leakage line is required.

3. Investigation of the hydraulic connections

This chapter introduces calculations for pump/motor leakage flow in order to identify a suitable connection for the external leakage line. The DDH system can be divided into three separated fluid volumes: the piston side of the cylinders A_1 and the rod side of the cylinders A_3 with the hydraulic accumulator B and the hydraulic accumulator A between the pump/motors. The accumulation of hydraulic fluid in any of these volumes will raise the pressure, as was shown in Figure 5 for the first prototype.

This chapter contains a simple model in which the volumetric efficiency of the pumps $\eta_{v,\,p}$, determines the actual volume produced and the leakage flow of the pump/motor. For simplicity, it is assumed that this efficiency is the same for both pumps/motors. This model is a rough simplification as in reality, the leakage flow is distributed both as internal and external leakage flow [25] and volumetric efficiency varies as a function of the differential pressure and speed. This case study aims to find out how leakages should be controlled so that the system is as insensitive to fluctuations in the volumetric efficiency as possible.

The calculations were performed by determining the change in the volume of fluid during the lifting and lowering with the maximum stroke length of the cylinder l [26]. The volume changes are determined during a single lifting–lowering cycle.

Next, the theoretical flow Q_{Vt}, the actual flow Q_{Vr} and the leakage flow Q_{Vl} for both pumps are determined. It is assumed that the speed of the electric motor is n in rev/s. In this case, the pumps produce theoretical flows of:

$$Q_{Vt1} = n \cdot D_1, \tag{1}$$

$$Q_{Vt2} = n \cdot D_2, \tag{2}$$

where D_1 and D_2 are the theoretical pump displacements for the pump/motors P_1 and P_2, respectively.

The actual flows Q_{Vr1} and Q_{Vr2} and leakage flows Q_{Vl1} and Q_{Vl2} in accordance with the volumetric pump efficiency η (D, p) are

$$Q_{Vr1} = \eta_{V,p} \cdot n \cdot D_1, \tag{3}$$

$$Q_{Vr2} = \eta_{V,p} \cdot n \cdot D_2, \tag{4}$$

$$Q_{Vl1} = \left(1 - \eta_{V,p}\right) \cdot n \cdot D_1, \tag{5}$$

$$Q_{Vl2} = \left(1 - \eta_{V,p}\right) \cdot n \cdot D_2. \tag{6}$$

During the lifting, the side flow $Q_{Vc1,up}$ of the cylinder A_1 is equal to the actual flow Q_{Vr1} produced by the pump/motor P_1. Theoretical flow Q_{Vt1} leaves the pump/motor P_1 during the lowering from A_3 cylinder's side. Therefore,

$$Q_{\mathrm{Vc1,up}} = Q_{\mathrm{Vr1}}. \tag{7}$$

$$Q_{\mathrm{Vc1,down}} = Q_{\mathrm{Vt1}}. \tag{8}$$

The cylinder losses are assumed to be negligible, so the required flows $Q_{\mathrm{Vc2,up}}$ and $Q_{\mathrm{Vc2,down}}$ are determined as

$$Q_{\mathrm{Vc2,up}} = R_{\mathrm{A}} \cdot Q_{\mathrm{Vr1}} = R_{\mathrm{A}} \cdot \eta_{\mathrm{V,p}} \cdot n \cdot D_1, \tag{9}$$

$$Q_{\mathrm{Vc2,down}} = R_{\mathrm{A}} \cdot Q_{\mathrm{Vt1}} = R_{\mathrm{A}} \cdot n \cdot D_1. \tag{10}$$

It is assumed that the cylinder is moved at a constant velocity v, which can be determined as follows for lifting and lowering, respectively:

$$v = \frac{Q_{\mathrm{Vs1}}}{A_1}, \text{ where } v_{\mathrm{up}} = \frac{Q_{\mathrm{Vr1}}}{A_1}, \ v_{\mathrm{down}} = \frac{Q_{\mathrm{Vt1}}}{A_1}. \tag{11}$$

In this case, the raising (t_{up}) and lowering (t_{down}) times can be calculated by combining Equations (1), (3) and (11):

$$t_{\mathrm{up}} = \frac{l}{v_{\mathrm{up}}} = \frac{l \cdot A_1}{Q_{\mathrm{Vr1}}} = \frac{l \cdot A_1}{\eta_{\mathrm{v,p}} \cdot n \cdot D_1}, \tag{12}$$

$$t_{\mathrm{down}} = \frac{l}{v_{\mathrm{down}}} = \frac{l \cdot A_1}{Q_{\mathrm{Vt1}}} = \frac{l \cdot A_1}{n \cdot D_1}. \tag{13}$$

As was stated earlier, the external leakage line of the pump/motors that are used should be connected to the line under 0.3 MPa. The next section will determine the structure. The least sensitive case will be selected as the system structure. The review will be carried out for a total of four different alternative cases. Figure 7a illustrates Case I, where both the leakage lines are connected to the line with hydraulic accumulator A. In Case II in Figure 7b, both external leakage lines are connected to hydraulic accumulator B. Figure 7c shows Case III, where the external leakage lines of the pump/motors P_2 and P_1 are connected to the line with hydraulic accumulators A and B, respectively. Figure 7d illustrates Case IV, where the external leakage lines of the pump/motors P_1 and P_2 are connected to the lines with hydraulic accumulators A and B, respectively.

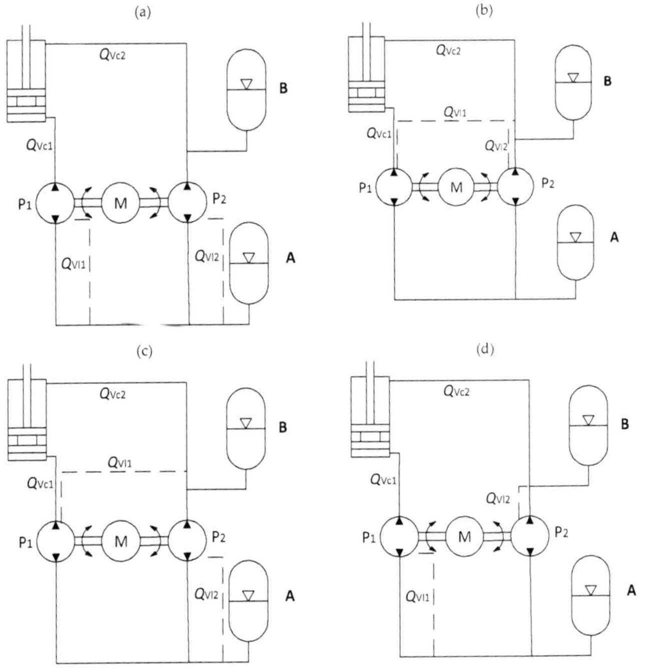

Figure 7. Cases to direct the pump/motor external leakage flows: (a) Case I, (b) Case II, (c) Case III and (d) Case IV.

3.1. Case I

In Case I, during the lifting, the A_3 side of the cylinder should remove flow equal to $Q_{Vc2,up}$, and the pump P_2 is able to remove the theoretical flow of Q_{Vt2}. Thus, during the entire lifting motion, the volume $\Delta V_{up,I}$ can be obtained with Equations (2), (9) and (12):

$$\Delta V_{up,I} = t_{up} \cdot \left(Q_{Vc2,up} - Q_{Vt2}\right) = l \cdot A_1 \left(R_A - \frac{R_Q}{\eta_{v,p}}\right). \tag{14}$$

During the lowering, the A_3 chamber of the cylinder should receive the flow of $Q_{Vc2,down}$. The pump/motor P_2 is able to produce the actual flow of Q_{Vr2}. Thus, during the whole cycle in the A_3 chamber of the cylinder, the following difference in volumes $\Delta V_{down,I}$ occurs, which is obtained with Equations (4), (10) and (13):

$$\Delta V_{down,I} = t_{down} \cdot \left(Q_{Vr2} - Q_{Vc2,down}\right) = l \cdot A_1 \left(\eta_{v,p} \cdot R_Q - R_A\right). \tag{15}$$

The entire cycle change of volume $\Delta V_{t,I}$ in Case I is:

$$\Delta V_{t,I} = \Delta V_{up,I} + \Delta V_{down,I} = l \cdot A_1 \cdot R_Q \left(\eta_{v,p} - \frac{1}{\eta_{v,p}} \right). \tag{16}$$

3.2. Case II

Figure 7b illustrates Case II, where both of the leakage lines are connected to the line with hydraulic accumulator B and A_3 chamber branch. During the lifting, the chamber A_3 should remove the flow $Q_{Vc2,\,up}$, as well as the leakage flows of the pumps Q_{Vl1} and Q_{Vl2}. The pump/motor P_2 is able to remove the theoretical flow Q_{Vt2}. In this case, the volume $\Delta V_{up,\,II}$ is obtained with Equations (2), (5), (6), (9) and (12):

$$\Delta V_{up,II} = t_{up} \cdot \left(Q_{Vc2,n} + Q_{Vl1} + Q_{Vl2} - Q_{Vt2} \right) = l \cdot A_1 \left(R_A + \frac{1-\eta_{v,p}}{\eta_{v,p}} - R_Q \right). \tag{17}$$

During the lowering, the A_1 chamber of the cylinder should remove the flow $Q_{Vc2,down}$. The pump/motor P_2 is able to provide the flow Q_{Vr2}. In addition, the same line will also have the external leakage flows Q_{Vl1} and Q_{Vl2}. In this case, the change in the volume $\Delta V_{down,\,II}$ can be obtained with equations (4), (5), (6), (10) and (13):

$$\Delta V_{down,II} = t_{down} \cdot \left(Q_{Vr2} + Q_{Vl1} + Q_{Vl2} - Q_{Vc2,down} \right) = l \cdot A_1 \left(R_Q + 1 - \eta_{v,p} - R_A \right). \tag{18}$$

The entire cycle change of volume $\Delta V_{t,II}$ in Case II is:

$$\Delta V_{t,II} = \Delta V_{up,II} + \Delta V_{down,II} = l \cdot A_1 \left(\frac{1}{\eta_{v,p}} - \eta_{v,p} \right). \tag{19}$$

3.3. Case III

In Case III, the leakage flow of the pump/motor P_2 is connected to the hydraulic accumulator A line and the P_1 leakage flow to the A_3 chamber line of the cylinder, as shown in Figure 7c. During the lifting, the A3 chamber of the cylinder should pump out the flow $Q_{Vc2,up}$ and the pump/motor P_1 the leakage flow Q_{Vl1}, but at the same time the pump/motor P_2 can remove the flow of Q_{Vt2}. Thus, during the whole stroke of the lift, the change in the volume is derived from Equations (2), (5), (9) and (12):

$$\Delta V_{up,III} = t_{up} \cdot \left(Q_{Vc2,up} + Q_{Vl2} - Q_{V2t} \right) = l \cdot A_1 \left(R_A + \frac{1-\eta_{v,p}}{\eta_{v,p}} - \frac{R_Q}{\eta_{v,p}} \right). \tag{20}$$

During the lowering, the A_3 chamber of the cylinder should provide the flow $Q_{Vc2,down}$. The pump/motor P_2 is able to pump out a flow equal to Q_{Vr2}. In this case, the total volume of $\Delta V_{down,III}$ is obtained with Equations (4), (5), (10) and (13):

$$\Delta V_{down,III} = t_{down} \cdot \left(Q_{Vr2} + Q_{Vl2} - Q_{Vc2,down}\right) = l \cdot A_1\left(\eta_{v,p} \cdot R_Q + 1 - \eta_{v,p} - R_A\right). \tag{21}$$

The entire cycle volume change $\Delta V_{t,\,III}$ in Case III is:

$$\Delta V_{t,III} = \Delta V_{up,III} + \Delta V_{down,III} = l \cdot A_1\left(R_Q - 1\right)\left(\frac{1}{\eta_{v,p}} - \eta_{v,p}\right). \tag{22}$$

3.4. Case IV

In Case IV, the leakage flow from the pump/motor P_1 is directed to the hydraulic accumulator A and the leakage flow of the pump/motor P_2 is directed to the hydraulic accumulator B. During the lifting, A_3 chamber flow of the cylinder $Q_{Vc2,up}$ and leakage flow Q_{Vl2} should be removed by the pump/motor P_2, but the pump/motor P_2 is able to actually pump out the flow of Q_{Vt2}. Thus, during the whole stroke, the change in the volume of the fluid chamber $\Delta V_{up,IV}$ is obtained with Equations (2), (6), (9) and (12):

$$\Delta V_{up,IV} = t_{up} \cdot \left(Q_{Vc2,up} + Q_{Vl2} - Q_{Vt2}\right) = l \cdot A_1\left(R_A - R_Q\right). \tag{23}$$

During the lowering, the A_3 chamber of the cylinder should pump out the $Q_{Vc2,down}$ volume flow. The required pumping flow rate of P_2 is Q_{Vr2} and the additional leakage flow is Q_{Vl2}. In this case, the change in the volume $\Delta V_{down,IV}$ is obtained with Equations (4), (6), (10) and (13):

$$\Delta V_{down,IV} = t_{down} \cdot \left(Q_{Vr2} + Q_{Vl2} - Q_{Vc2,down}\right) = l \cdot A_1\left(R_Q - R_A\right). \tag{24}$$

The entire cycle change in the volume $\Delta V_{t,\,IV}$ in Case IV is:

$$\Delta V_{t,IV} = \Delta V_{up,IV} + \Delta V_{down,IV} = 0\ . \tag{25}$$

4. Summary

Four cases of the locations of pump/motor leakages were investigated. The different implementation variants were determined by the changes in the volume of the cylinder as a function of the volumetric efficiency, and the results are shown in Figure 8. As a result, Case IV appears to be the most independent case [26]. However, this method did not investigate the pressure built up with various payloads. Neither did this investigation consider the practical aspects of realizing such a system. Therefore, the system structure will be investigated further in Matlab Simulink.

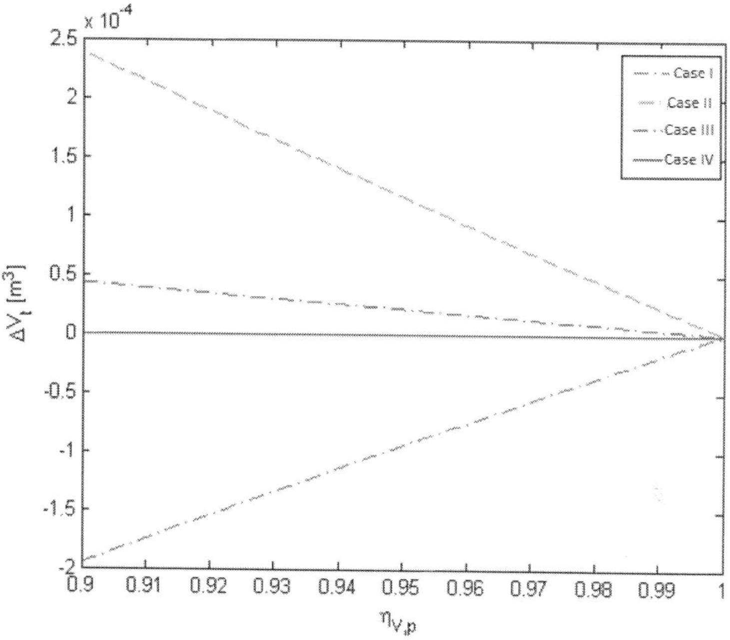

Figure 8. Results of the DDH system with four cases of external leakage connections [26].

5. Simulink modeling

Matlab Simulink R2015a and Simscape SimHydraulics and the SimPowerSystems component library were used for creating the simulation model. The following assumptions were made:

- The effect of temperature on fluid properties is neglected.
- External leakages in the hydraulic pump/motor are neglected. Volumetric efficiency is assumed to be 1.
- Losses in the hydraulic accumulator are ignored.

The simulation model for the test setup (introduced in Figure 2) is shown in Figure 9.

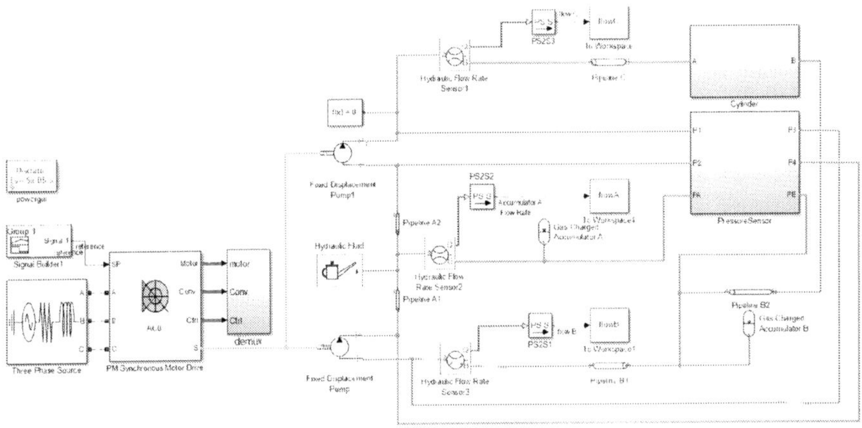

Figure 9. Simulation model of the DDH system.

Figure 10 illustrates the simulation results: reference signal, motor speed and cylinder position. A maximum payload of 150 kg was utilized for the simulation.

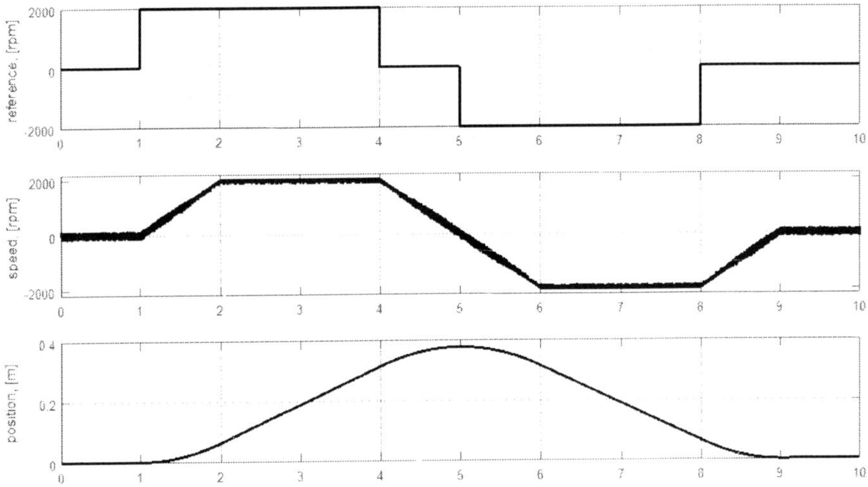

Figure 10. Simulation results of the DDH system: reference signal, motor speed and cylinder position.

The movement of the cylinder is smooth and identical to the first prototype demonstrated in Figure 5. The flows in volume A (the line with accumulator A), B (the line with accumulator B) and C (line between the pump/motor P_1 and the cylinder) are illustrated in Figure 11.

http://dx.doi.org/10.5772/61793

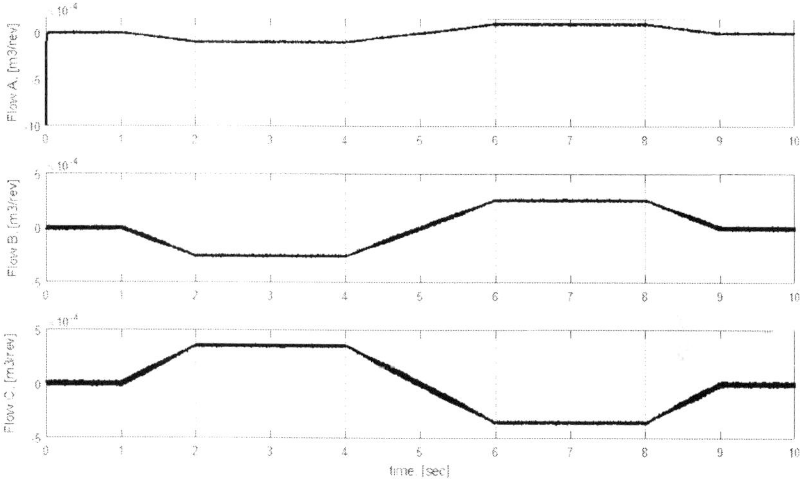

Figure 11. Simulation results of the DDH system: flow in lines A, B and C.

Flows B and C are mirrors of each other, which corresponds to correct behavior. Conversely, flow A has an initial drop which corresponds to sucking in oil from accumulator A during the initial lifting motion.

Figure 12 illustrates the experimental results for two system configurations with hydraulic accumulators: A and A + B. The pressure with accumulator A (red line) configuration goes up to 2.7 MPa, whereas the proposed combination A + B maintains the pressure in the range of 0.7 MPa.

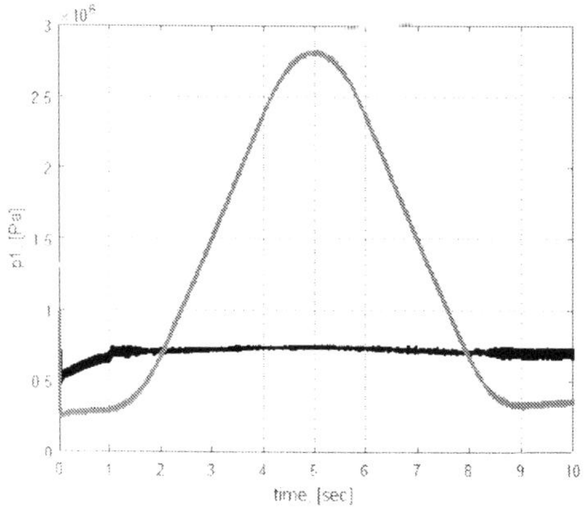

Figure 12. Simulation results of the pressure in the DDH system only with accumulator A (red line) and with accumulators A + B (black line)

The system pressure is illustrated in Figure 13, where p_1 is the P_1 pump/motor pressure, p_3 is the P_2 pump/motor pressure, and p_2, p_4 are the "tank" line pressures (line with accumulator A).

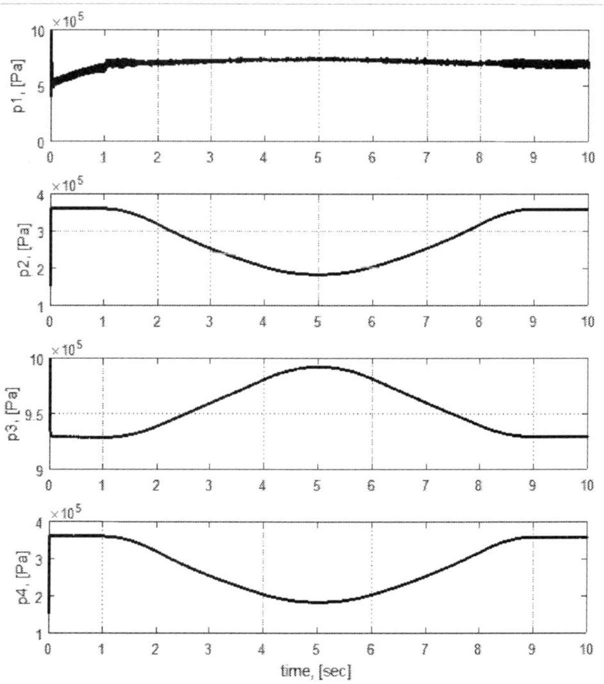

Figure 13. Simulation results of the pressure in the DDH system: p_1 is the P_1 pump/motor pressure, p_3 is the P_2 pump/motor pressure, and p_2, p_4 are the "tank" line pressures (accumulator A).

The pressure p_3 ranges between 0.93 MPa and 1 MPa during the whole lifting–lowering cycle with a maximum payload of 150 kg. The tank pressure (p_2 and p_4) varies between 0.38 MPa and 0.18 MPa. Thus, according to the simulation, Cases II, III and IV do not fulfill the requirements of the leakage line where the maximum allowed constant pressure is 0.3 MPa and in the short term, it is 1 MPa. According to the results shown in Figure 13, from the proposed connection cases in Figure 7, only Case I can be used for the realization of the DDH setup, where both external leakage lines are connected to line A (accumulator A).

6. Experimental investigation

The essential parameters, such as currents, voltage, pressure, flow and height, were measured from the DDH setup. The location of the utilized sensors is illustrated in Figure 4. The experimental efficiencies and energies of the DDH test setup system for lifting movements were calculated as shown below:

$$\eta_{\text{up\_tot}} = \frac{E_{\text{pot}}}{E_{\text{mech}}}, \tag{26}$$

where E_{pot} is the potential energy of the payload

$$E_{\text{pot}} = mgH, \tag{27}$$

where m is the mass of the payload in kg and g is a gravitational constant in m/s^2. H is the position of the cylinder piston in m.

E_{mech} is the energy of the shaft and is calculated as the integral of the power at the shaft (P_{shaft}).

$$P_{\text{shaft}} = T\Omega, \tag{28}$$

where P_{shaft} is the output energy of the shaft in W, T is the motor torque in Nm and Ω is the angular speed in rad/s. A motor control algorithm was utilized to measure the angular speed and estimate the motor torque.

The output energy of the hydraulic part E_{hydr} is calculated as the integral of the output hydraulic power:

$$P_{\text{hydr}} = pv_c A, \tag{29}$$

where p is the pressure in Pa, v_c is the velocity of the cylinder piston in m/s and A is the cross area of the cylinder piston in m^2.

Depending on the operating point of a hydraulic pump/motor unit in its performance curve, the relationship between the flow and hydraulic losses in a system varies significantly. During the lifting, the hydraulic pump/motor unit operates as a pump. The input energy of the pump is mechanical energy and the output is hydraulic energy. Figure 14 presents an example of a Sankey diagram for the measured losses and efficiencies of the DDH system.

In Figure 14, hydro-mechanical losses equal 33.4%. Hydro-mechanical losses include shaft and hydraulic losses. Hydraulic losses in DDH systems are composed of pipe friction losses and other fittings, entrance and exit losses and losses from changes in the pipe size resulting from a reduction in the diameter, pump/motors and cylinder losses. The overall cylinder efficiency is mostly dependent on the frictional losses encountered by the piston and the rod during its stroke. Frictional losses depend on the pressure difference across the seal, sliding velocity, seal material, temperature, time, wear and direction of the movement. Measured electrical machine

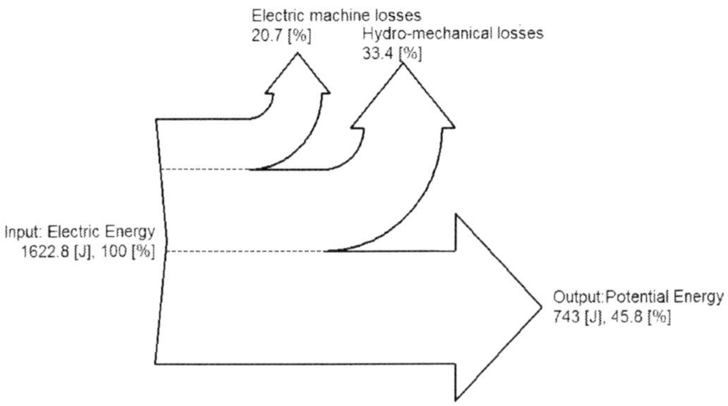

Figure 14. Measured Sankey diagram of the DDH with a motor speed of 400 rpm and a payload of 150 kg (the efficiency of the frequency converter is not included) [19].

losses are 20.7%. Electrical machine losses are composed of the following elements: stator and rotor resistive losses, iron losses, additional losses and mechanical losses. Mechanical losses include friction in the motor bearings. Bearing losses depend on the shaft speed, bearing type, properties of the lubricants and the load. The converter losses take place mostly in the semiconductor switches and in the auxiliary power systems. In this study, the losses of the frequency converter are not included.

It was challenging to create the DDH test setup due to the asymmetrical double-acting cylinder used and the difficulty of finding matching displacements of the pump/motors in order to fulfill $R_Q = R_A$. Moreover, another challenge was the location of the hydraulic accumulators and external leakage pump/motor lines. The Sankey diagrams showed that electrical machine losses are in second place after the dominant hydro-mechanical losses. Taking pressure balancing into consideration, the hydro-mechanical losses in the system will be reduced in the second DDH prototype.

7. Conclusion

This paper described a DDH setup and its potential for applicability to NRMMs. It investigates the compensation of pump/motor displacement for an asymmetrical double-acting cylinder and the location of hydraulic accumulators and external leakage pump/motor lines. The review was carried out for four different alternative cases. The mathematical model used in the review suggested Case IV as the least sensitive case. Further investigations were performed in Matlab Simulink. According to the simulations, Cases II, III and IV did not fulfill the requirements of a leakage line where the maximum allowed constant pressure was 0.3 MPa and, in the short term, it was 1 MPa. Therefore, only Case I can be used for the realization of the DDH setup, in which both external leakage lines are connected to line A (accumulator A).

The experimental tests demonstrated that direct-driven hydraulics (DDH) has the advantage of a fully self-contained electrohydraulic actuator, which combines the high power density of hydraulics and the accuracy of an electric motor. The measured energy efficiency of the DDH varies by up to 46% with the direction of the motion of the cylinder and the motor speed. As the Sankey diagram showed, the hydro-mechanical losses dominate in the original DDH setup. As expected, with regard to the efficiency of the DDH setup, the weak link in the chain is found in the losses of the hydro-mechanical components of the system. Therefore, further studies are required on the improvement of DDH.

Acknowledgements

The research was enabled by the financial support of ArcticWell project (Academy of Finland) and internal funding at the Department of Engineering Design and Production at Aalto University.

Author details

Tatiana A. Minav[1*], Jani E. Heikkinen[2] and Matti Pietola[1]

*Address all correspondence to: tatiana.minav@aalto.fi

1 Department of Engineering Design and Production, School of Engineering, Aalto University, Finland

2 Independent researcher, http://orcid.org/0000-0002-5991-7580, Helsinki, Finland

References

[1] Tier 4 Emission Standards for Nonroad Diesel Engines, online, https://www.dieselnet.com/standards/us/nonroad.php#tier4.

[2] Liukkonen, M., Lajunen, A., Suomela, J. Comparison of different buffering topologies in FC-hybrid non-road mobile machineries. At the 7th IEEE Vehicle Power and Propulsion Conference, Chicago, September 6–9, 2011. Chicago 2011.

[3] Wang, T. et al. An energy-saving pressure-compensated hydraulic system with electrical approach. ieee/asme transactions on mechatronics, vol. 19, no. 2, april 2014 Mech., 2013.

[4] Bifeng, Y., Tao, G., Shengji, L.,. Investigate on the combustion and emission characteristics of small non-road diesel engine fueled with bio-diesel, In 2012 (ISDEA), 2012, pp. 852–856.

[5] Skinner, J., Smith, A., Frischemeoer, S., Holland, M. Advancements in hydraulic systems for more electric aircraft. Proceedings of MEA 2015 Conference, Toulouse, France, February 2015.

[6] Youzhe, J., Song, P., Li, G., Zhanlin, W., Lihua, Q. Pressure loop control of pump and valve combined EHA based on FFIM. The Ninth International Conference on Electronic Measurement & Instruments, 2009.

[7] Zhang, Q., Li, B. Feedback linearization PID control for electro-hydrostatic actuators, at 2nd International Conference on Artificial Intelligence, Management Science and Electronic Commerce (AIMSEC), 2011, p. 358 - 361, DOI: 10.1109/AIMSEC.2011.6010249

[8] Liang, B., Li, Y., Zhang, Z. Research on Simulation of Aircraft Electro-Hydrostatic Actuator Anti-Skid Braking System (ICMTMA), 2011.

[9] Altare, G., Vacca, A., Richter, C. A novel pump design for an efficient and compact electro-hydraulic actuator. IEEE Aerospace Conference, 2014 IEEE, 2014, pp. 1–12.

[10] Ahn K. K., Nam D.N.C.., Jin M.,. Adaptive backstepping control of an electrohydraulic actuator. IEEE/ASME Transactions on Mechatronics, Vol. 19, No. 3, June 2014.

[11] Mini-Motion Package (MMP), [online] http://www.kybfluidpower.com/Mini_Motion_Package.html.

[12] AC Servo Motor Driven Hydraulic Pump Control System, [online] http://www.yuken.co.uk/long_cat/%5BK%5D/816-817.pdf.

[13] Daher, N., Ivantysynova, M. Electro-hydraulic energy-saving power steering systems of the future. Proceedings of the 7th FPNI PhD Symposium on Fluid Power, 27-30 June 2012, Reggio Emilia, Italy.

[14] Michel, S., Weber, J., Electrohydraulic compact-drives for low power applications considering energy-efficiency and high inertial loads. 2012. Fluid Power and Motion Control FPMC 2012 conference, Bath, UK, p.93-109

[15] Michel, S., et al. Energy-efficiency and thermo energetic behavior of electrohydraulic compact drives, at 9th International Fluid Power Conference (IFK), 24 - 26 March 2014, Aachen, Germany.

[16] Busquets, E., Ivantysynova M. Temperature prediction of displacement controlled multi-actuator machines. Int. J. of Fluid Power, Jan., 2014.

[17] Busquets, E. An investigation of the cooling power requirements for displacement-controlled multi-actuator machines, M.Sc thesis, Purdue university, 2013.

[18] Minav, T., Bonato, C., Sainio, P., Pietola, M. Direct driven hydraulic drive, at 9th International Fluid Power Conference (IFK), 24 - 26 March 2014, Aachen, Germany.

[19] Minav, T.A., Bonato, C., Sainio, P., Pietola, M. Efficiency of direct driven hydraulic drive for non-road mobile working machines, at 2014 International Conference on Electrical Machines (ICEM), 2-5 September 2014, Berlin, Germany, pp. 2431–2435.

[20] Vivoil motor, Data Sheet: reversible motor - series XV, http://www.vivoil.com/files/xm_en/xm201.pdf, visited on September 8, 2013.

[21] Emerson Control Techniques Unidrive SP1406 drive, http://www.emersonindustrial.com, visited on September 8, 2013.

[22] Emerson Control Techniques Unimotor 115U 2C, http://www.emersonindustrial.com/en-en/documentcenter/ControlTechniques/Brochures/unimotor_fm_product_data.pdf, visited on 1 September, 2013.

[23] Bosch Rexroth Indradyn T Torque motors. [online]. http://www.boschrexroth.com/dcc/content/internet/en/pdf/PDF_p146807_en.pdf.

[24] Bosch Rexroth AZMF External gear motors. [online]. www.boschrexroth.com.

[25] Kauranne, H., Kajaste, J., Vilenius, M. Hydraulitekniikka, Helsinki: Sanoma Pro Oy, 2013. 496 p. ISBN 978-962-63-0707-7.

[26] Kiesi, M. Suorasähkökäyttöisen hydraulijärjestelmän paineakun valinta ja mitoitus, Bachelor Thesis, Aalto University, 2014.

Chapter 7

Variable Frequency Drive Applications in HVAC Systems

Yunhua Li

Additional information is available at the end of the chapter

http://dx.doi.org/10.5772/61782

Abstract

Building heating ventilation and air-conditioning (HVAC) systems are designed to operate at the peak load, which only occurs in a very short period of time throughout the year. One of the most effective ways to improve building energy efficiency is to utilize the variable frequency drives (VFDs). They are widely used in the HVAC field, including fans, pumps, compressors, etc. In a VFD-equipped system, the VFD adjusts the speed of one or more motors based on the system load requirements and operation schedule, resulting in a dramatic cut in energy consumption.

This article systematically introduces the application of VFDs in HVAC systems, including heating systems, ventilation systems, cooling systems, and refrigeration systems in commercial and industrial buildings. First, the principles are introduced for VFDs, variable speed fans, pumps, and compressors. Next, the control application of VFDs on each type of system (equipment) is summarized. In this section, the schematic diagram of each system as well as the detailed control strategy for each application is presented. In addition, the application of VFD on fault detection and diagnosis (FDD) is introduced. Lastly, some technical issues and concerns are discussed with potential solutions provided.

Keywords: Variable frequency drives, HVAC systems, control, energy efficiency, savings

1. Introduction

Electric motor-driven systems (EMDS) are the largest electrical end users and account for 43–46% of all global electricity consumption [1]. In the United States, HVAC and refrigeration applications consume 91% of motor-driven energy use in the residential sector and 93% in the commercial sector [2]. Reducing unnecessary energy consumption is the most direct and effective way to improve building energy efficiency. The increasing advancement of electronic and control technology greatly improves the performance of variable speed drives. Variable frequency drives (VFDs) are being used in the HVAC industry more frequently and in more applications. They can modulate the motor speed very smoothly within a wide range. Reduced motor speed provides a significant reduction in motor power.

In general, a VFD can be used in a variety of applications where the load is variable, such as fans, pumps, and compressors. The energy savings for fans and pumps are about 30–50% over conventional speed applications, and up to 35% for compressors. In the following sections, the working principle of the VFD is introduced and its applications on HVAC systems are presented. Lastly, the current issues and concerns about VFDs are discussed with potential solutions provided.

2. Principles

2.1. VFD

A VFD is an electrical device used to control the rotation speed of an alternating current (AC) electric motor by adjusting the frequency of the electrical power supplied to the motor. It is also called pulse-width-modulated drive (PWM drive), adjustable-speed drive (ASD), adjustable-frequency drive (AFD), AC drive, inverter drive [3], or variable voltage variable frequency (VVVF) drive.

Most VFDs used in HVAC applications are inverters using sine-coded PWM technology. As shown in Figure 1, a VFD works by converting the incoming AC power to DC power using a diode bridge rectifier, then passing the filtered, smoothed the voltage onto the inverting section, and finally controlling voltage and frequency sent to the motor by high-speed bipolar transistors [4].

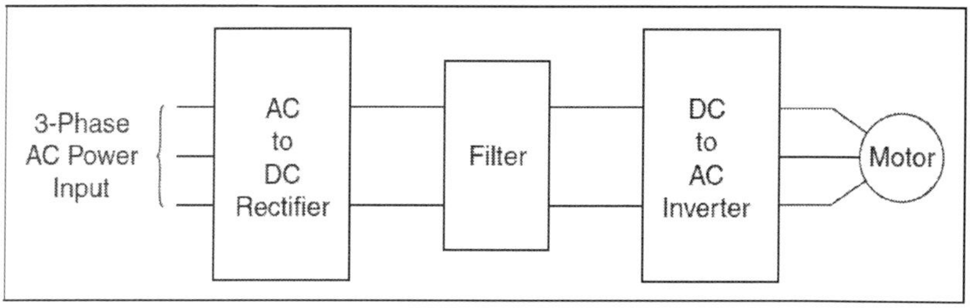

Figure 1. Schematic diagram of a variable frequency drive [4]

The output voltage is adjusted by changing the width and number of the voltage pulse as shown in Figure 2, whereas the output frequency is varied by changing the length of the cycle.

http://dx.doi.org/10.5772/61782

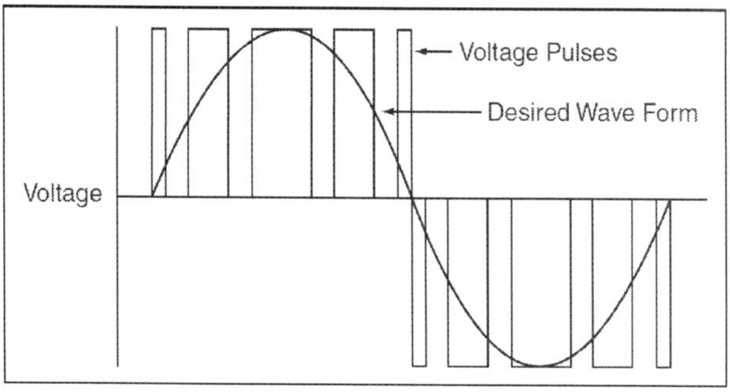

Figure 2. Sine-coded PWM waveform [4]

A VFD can provide the follow benefits:

- Soft-start capability reduces the inrush current when motors start up, and thereby reduces the mechanical stress on the motor and improves the motor reliability.
- Step-less regulation for motor speed electrically.
- Reduces the motor power consumption significantly with proper controls.
- Improves the power factor of the entire drive system including VFD and motors.
- Metering and monitoring system performances.

These advantages boost the widespread utilization of VFDs in a variety of industrial processes including the HVAC field. Although equipping a VFD in a new system or existing system increases the initial investment, the reduced VFD cost combined with the increased energy savings derived from a VFD result in short payback period, which is usually less than three years.

2.2. Variable speed fans and pumps

Variable speed fans and pumps are the fans and pumps equipped with VFDs. Since their speeds vary as the frequency of supply power changes, these pumps and fans are also called variable frequency fans and pumps. The operating characteristic of centrifugal fans and pumps makes them the excellent candidates for VFD applications. According to the fan and pump affinity laws, the fan or pump power has a cubic relationship with the motor speed. Therefore, significant power savings could be achieved by reducing the motor speeds with the proper controls.

The variables related to the fan or pump performance are impeller diameter D, rotation speed N, gas/water density ρ, volume flow rate Q, pressure P, power W, and mechanical efficiency η. In a typical application, the fan or pump diameter is constant. The airflow or water flow, fan or pump head, and power solely relies on the speed. These relations are presented by the following equations [3] (where equations (1) and (2) denote two working conditions):

$$\frac{Q_1}{Q_2} = \frac{N_1}{N_2} \tag{1}$$

$$\frac{P_1}{P_2} = \left(\frac{N_1}{N_2}\right)^2 \tag{2}$$

$$\frac{W_1}{W_2} = \left(\frac{N_1}{N_2}\right)^3 \tag{3}$$

Equation (3) clearly indicates how the speed change affects the power change. For example, 50% speed reduction results in 87.5% power reduction theoretically. There are several types of air and water systems including the single-duct variable air volume (SDVAV) system, dual-duct variable air volume (DDVAV) system, single-zone system, multi-zone system, primary and/or secondary chilled water system, and the hot water system. Each type of system requires specific control strategy rather than a fixed low-speed operation without modulation.

2.3. Variable speed compressors

Variable speed compressors or variable frequency compressors are compressors equipped with VFDs. In the HVAC industry, there are several types of compressors associated with the refrigeration systems: reciprocating, screw, scroll, and centrifugal. All of them are viable components for a VFD application. The commonly used working media includes air and refrigerant. Air compressors are frequently used in industrial processes as a power source for pneumatic control systems. Refrigerant compressors are typically used in air conditioners, unitary air-handling units (AHU), or chillers in residential and commercial buildings. Significant progresses on the theoretical study and applications on variable speed compressors were made during the past thirty years. As early as 1982, Itami et al. [5] conducted an experimental study on the performance and reliability of a rotary compressor and reciprocating compressor combined with frequency inverters. In 1996, Qureshi and Tassou [6] presented a review of application of variable speed capacity control on refrigeration systems. It pointed out that VFD applications on compressors were largely restricted to small-capacity air-conditioning units thus far, with only a small amount of applications to medium-range capacity units. More research work was still required. Since then, dramatic studies were conducted on applications of variable speed drives in HVAC systems [7–9].

Unlike other types of compressors, centrifugal compressors have similar energy performance as centrifugal fans and pumps. A scroll compressor is especially suitable for a VFD due to its intrinsic structural design. Scroll compressors and reciprocating compressors are the most commonly used types on rooftop units (RTUs) and heat pumps (HPs) units. For the majority of existing constant speed compressors, installing a VFD on the compressor is the most feasible way. In recent years, some original equipment manufacturers (OEMs) of compressors, such as Emerson and Danfoss, have already manufactured variable speed compressors. Several large HVAC manufacturers, such as Trane, Carrier, McQuay, Lennox, AAON, York, and Emerson, started equipping variable speed compressors in their new products, such as RTUs, HPs, computer room air-conditioning (CRAC) units, or chillers.

Unlike fans and pumps, the scroll and reciprocating compressors usually do not have a cubic relationship between power and frequency. Figure 3 shows an approximate linear relationship between power and frequency for a 5-ton RTU with a scroll compressor, based on the experimental study [10]. This relationship clearly demonstrates how the compressor power changes with frequency. The linear correlation is also helpful to develop a new, simpler compressor model [11].

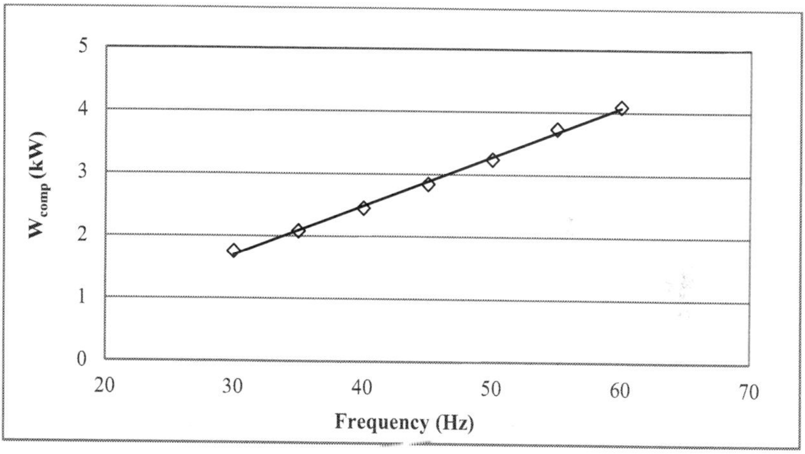

Figure 3. Relationship between compressor power and frequency for a 5-ton RTU

3. VFD applications

3.1. Air systems

3.1.1. Single-zone Variable Air Volume (VAV) system

Single-zone VAV system is the simplest air system. The VAV system mainly includes outdoor air and return air dampers, filter, heating and cooling coils, and a supply fan. Some units may also have a preheat coil, bypass damper, and return fan. Figure 4 presents a typical single-zone VAV system.

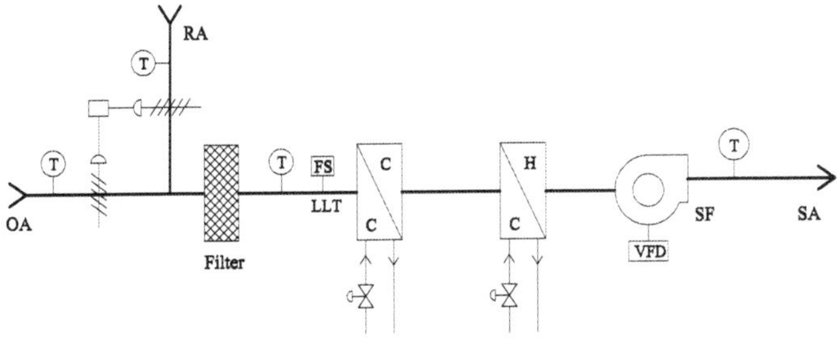

Figure 4. Typical single-zone VAV system

Typically, a single-zone air-handling unit is operated to control the temperature of only one space. The conventional method is to integrate the cooling and heating valve to control the space cooling and heating temperature set point. For a VFD-equipped supply fan, the fan speed can be modulated to maintain the space temperature set point, whereas the cooling and heating coil valves are used to control the supply air temperature (SAT).

Since the 2010 edition of ASHRAE Standard 90.1 [12], some requirements were added for single-zone VAV system control. It required that single-zone AHU and fan coil units with chilled water cooling coil and supply fan with motor greater than 5 hp shall have supply fans controlled by two-speed motors or VFD. Similarly, all the AHUs and AC units with direct-expansion (DX) cooling coil and a capacity ≥110,000 Btu/h that serve single zones shall have their supply fans controlled by the two-speed motor or VFD. These requirements are mandatory.

There are many successful case studies on VFD applications for single-zone units in projects. Li et al. [13] applied the VFD technology to a dozen single-zone systems by installing a VFD on each of them. They demonstrated that installing VFDs on supply fans in a series of single-zone units can save much more energy than running half of the units at constant speeds and shutting off the remaining half.

3.1.2. Single-duct VAV system

The single-duct VAV system is the most popular system, which comprises a main AHU, ductwork and a number of terminal boxes. The air-handling units are comprised of an outdoor air damper and return air damper, filter, preheating coil, cooling coil, and safety devices. Figure 5 shows a typical SDVAV system.

In a single-duct VAV system, VFDs are installed on the supply fan and return fan. Typically, the supply fan speed is modulated to maintain the duct static pressure at its set point. As the system load is reduced, the VFD speed decreases to maintain the same set point. Meanwhile, the set point need not be kept at a constant value. With the system load reduced, less airflow is needed to deliver to the space. The static pressure set point could be reset to meet the condition. This set point can be reset based on the VFD speed or supply fan airflow [14].

http://dx.doi.org/10.5772/61782

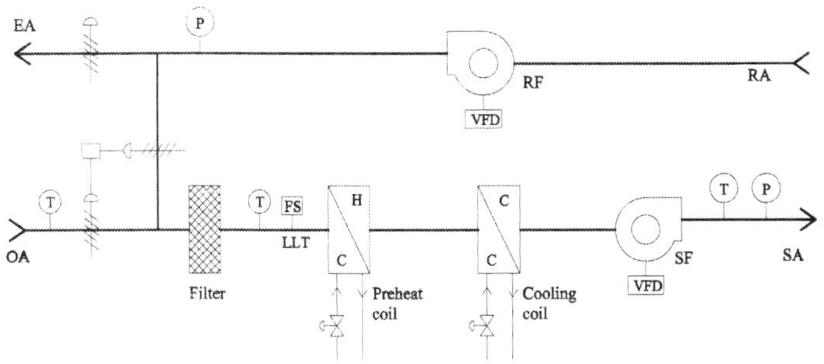

Figure 5. Typical single-duct VAV system

For the return fan, there are several control methods: modulating the return fan speed to maintain (a) return duct static pressure or (b) building differential pressure. However, these controls are not reliable due to the pressure measurement. A new control method is using a volume tracking method to maintain the airflow difference between the supply and return fans.

3.1.3. Dual-duct VAV system

A dual-duct variable air volume (DDVAV) system handles hot and cold air separately and delivers them through hot and cold ductwork. The hot air and cold air are mixed at the terminal box and then supplied to the space. There are two types of DD system: the single-fan dual-duct system and the dual-fan dual-duct system. The first one has a supply fan delivering the airflow to both hot and cold decks. The second one has a dedicated supply fan in each deck. The cold deck includes a cooling coil, whereas the hot deck is equipped with a hot water or steam coil. Figure 6 shows the schematic diagram of a single-fan DDVAV system.

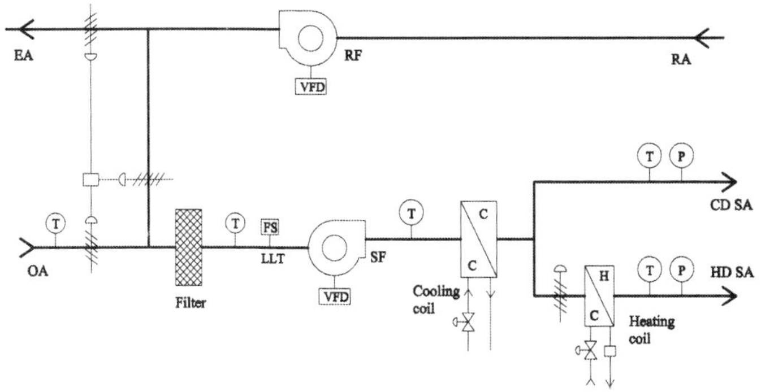

Figure 6. Single-fan DDVAV system schematic diagram

In a single-fan dual-duct VAV system, a VFD is installed on the supply fan. For a dual-fan dual-duct VAV system with separate supply fans for the hot and cold deck, a VFD is installed on each fan. If there is a return fan in this system as well, a VFD is also equipped on the return fan.

Typically, for a single-fan dual-duct system, the supply fan is modulated to maintain the cold deck static pressure, whereas the hot deck main damper is modulated to maintain the hot deck static pressure set point. For a dual-fan dual-duct system, each supply fan speed is modulated to maintain its own static pressure set point. Similarly, with single-duct VAV system, the return fan speed is modulated to maintain the airflow difference between the supply and return fans.

The energy savings of a dual-duct VAV system are often derived from the fan speed control and duct supply air temperature reset. Liu and Claridge [15] presented the models for the maximum potential energy savings by optimizing the hot deck and cold deck reset schedules, where 75% in potential savings can be expected.

3.1.4. Multi-zone system

A multi-zone system serves multiple zones with each zone having its own thermal requirement. Like a dual-duct system, one multi-zone system has cold and hot decks. However, the difference is that the cold air and hot air mixes at the outlet of air-handling unit before delivery to the space, whereas in a dual-duct system the hot air and cold mixes at the terminal boxes. Figure 7 shows the schematic diagram of a typical multi-zone system where a VFD is installed on the supply fan.

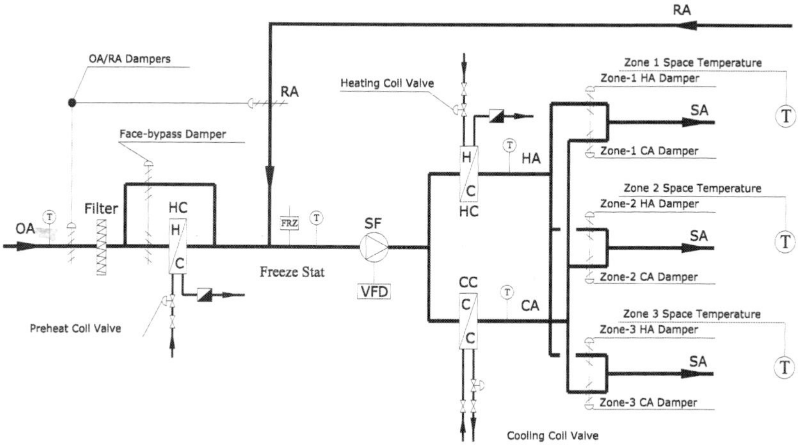

Figure 7. Multi-zone VAV system (three zones)

In a multi-zone system, the supply fan speed is modulated to maintain the discharge air static pressure or the temperature in the worst zone at its set point. The zone damper is modulated to maintain each zone temperature set point.

3.1.5. Exhaust air system

An exhaust air system is often associated with one air-handling unit, make-up unit, or fresh air unit. An exhaust air system is applicable for several types of facilities, such as kitchens, cafeterias, and laboratories in the hospital, just to list a few. They require enough fresh air and associated exhaust air. Proper exhaust airflow should be provided to satisfy the building or space pressure requirement. As the airflow delivered by air-handling unit is variable, the exhaust airflow is adjustable accordingly. Figure 8 shows an exhaust air system where a VFD is installed on the exhaust fan.

In this exhaust air system, the VFD is modulated to maintain the suction air pressure set point, or the differential airflow between the supply and exhaust air to maintain the required building pressure.

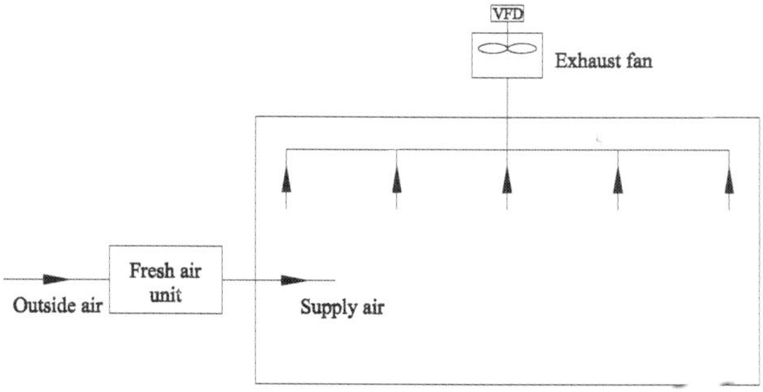

Figure 8. Exhaust air system

3.2. Water systems

The major water systems in HVAC system include chilled water system, condenser water system, and hot water system. Each system has dedicated pumps circulating water through a closed or an open loop. VFDs can be installed on these systems, which could reduce the pump energy consumption at partial load conditions.

3.2.1. Chilled water system and condenser water system

Chilled water system and condenser water system are two independent systems in the chiller plant. Figure 9 shows a typical chiller plant comprising these two loops. A chilled water system includes one or more chillers, chilled water pumps, and cooling coils. The cooling coils are usually located in the AHUs or fan coil units. There are two types of pumping system: primary-only system and primary–secondary system. In a primary-only system, the chilled water pump circulates the chilled water through the evaporator of chillers and cooling coils. In a primary–secondary system, there are two loops. The primary pumps circulate chilled water through the chiller only, while the secondary pumps circulate the chilled water through buildings. Usually, there is one bypass pipe, which connects the primary and secondary water loops. Many investigations and case studies were conducted on the efficiency, reliability, and optimization of primary–secondary or primary-only chilled water systems [16–18]. When VFDs are installed on chilled water pumps, how to operate pumps under the maximum efficiency point for single or multiple pumps are one of the study topics.

In a chilled water system, as seen in Figure 9, the cooling load of each coil varies at different zones and times, making the required chilled water flow variable. The primary pumps are modulated to maintain the loop differential pressure while simultaneously maintaining the minimum water flow requirement for chillers. The secondary pump speeds are equal to the primary pump speeds. As the building cooling load reduces, the required chilled water flow decreases. Reduced pump flow results in great pump power savings.

In a condensing water system, the condensing water pump circulates the condensing water through the condenser of chillers and cooling tower. When a VFD is installed on the condensing water pump, the pump speed is adjusted to maintain the loop differential pressure (ΔP) or temperature difference (ΔT).

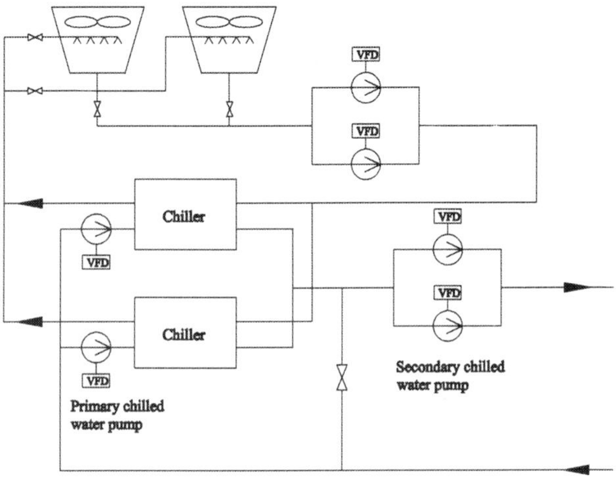

Figure 9. Chilled water and condenser water system

Furthermore, the VFDs could be installed on the fans of cooling tower. The fan speed is optimized to maintain the condensing water leaving temperature from the cooling tower.

http://dx.doi.org/10.5772/61782

3.2.2. Hot water system

The hot water system delivers the hot water from boilers or heat exchangers to the heating coils of air-handling units or terminal boxes inside the building. In traditional operation, the water pumps are running at full speed. The heating valves at the end users are modulated to control the airside temperature set point. Figure 10 shows a hot water system with VFD installed on both primary and secondary pumps. After installing of VFDs, the speed of secondary pump is often modulated to maintain the supply and return temperature difference or loop differential pressure. The speed of primary pump can track that of secondary pump and should be high enough to ensure sufficient water going through boilers.

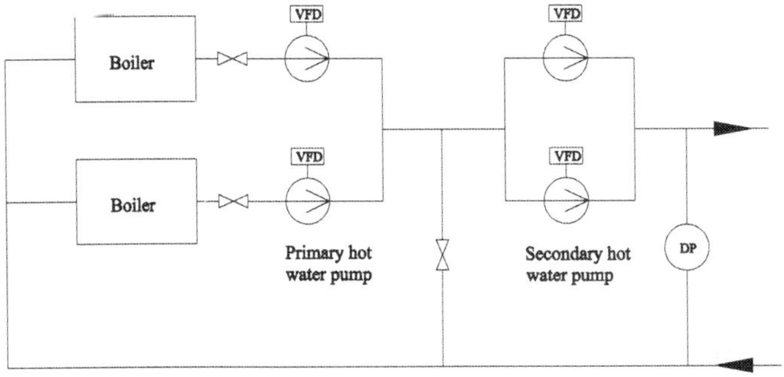

Figure 10. Hot water system

3.3. Air compressors

Compressed air has many applications in the manufacturing process. In the HVAC industry, air compressors can be used to generate the pressurized air to drive the pneumatic actuators for dampers and valves in air-handling units. The compressed air is stored in a pressurized tank, which serves as an air source to the end users. Traditionally, the pressure of tank is maintained by the on–off control of one or multiple air compressors. Figure 11 shows a schematic diagram of an air compressor system with a VFD installed on each compressor.

Typically, staging control is used to maintain the compressed air pressure. When the end users require less compressed air and the compressed air pressure is higher than the set point, the compressor will shut off. On the contrary, one more compressor starts when the end user utilizes more compressed air and the compressed air pressure drops down below the set point. This inefficient control causes frequent compressor start–stops, which definitely shortens the lifetime of the compressor. However, if a VFD is installed, the wear and tear on the compressors is less so that their lifetime is prolonged. In addition, the compressor power is reduced.

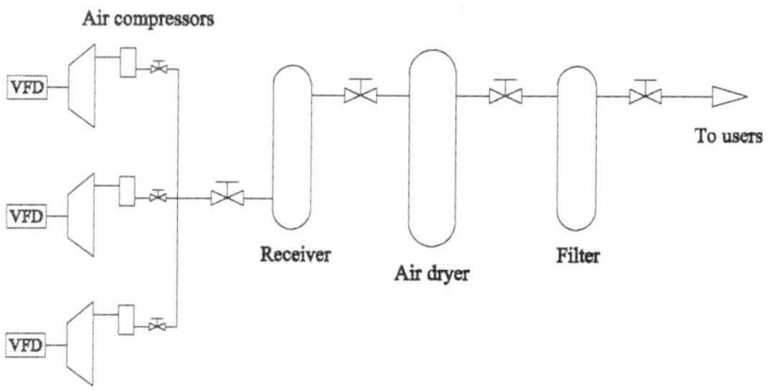

Figure 11. Air compressor system

3.4. Refrigeration systems

Refrigeration systems are also good candidates for VFD applications. The compressor is the major device where the VFD is installed in a refrigeration loop. The typical applications include RTUs, HPs, CRAC units, and chillers.

3.4.1. *Rooftop units*

Rooftop unit is one type of unitary air handler designed for outdoor use, usually on the roof. There are two types of configurations: packaged unit and split unit. A typical packaged rooftop unit has a refrigeration system delivering cool air into the space. Therefore, it is also called the direct-expansion (DX) unit. Meanwhile, most RTUs provide heating to the space using either a gas heater or an electric heater.

In an RTU, the supply fan (or indoor fan) and compressors are usually running at a constant speed. At partial load conditions, excessive fan and compressor power are consumed due to constant speed operations. With the installation of VFDs on the fan and (or) compressors, significant power consumption could be saved. In addition, compressors account for the largest part of power consumption in an RTU. With the reduction of compressor speed, both the demand and energy consumption are reduced greatly.

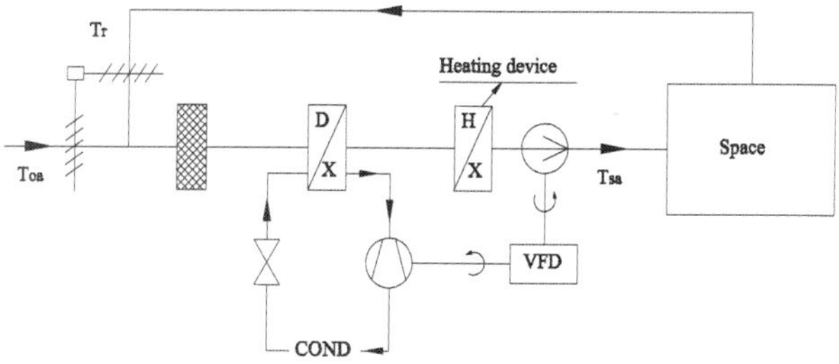

Figure 12. Single-stage DX rooftop unit

http://dx.doi.org/10.5772/61782

Figure 12 is a typical single-stage RTU. The supply fan circulates air through the evaporator and heater. Traditionally, the supply fan and compressor run at a constant speed. Initially, the VFD is used on the supply fan to modulate the fan speed and maintain the space temperature set point. This type of RTU can be called variable capacity RTU. Later, the VFD is used on the compressor as well. Both fan and compressor speeds can be modulated to control the space temperature.

3.4.2. Heat pump units

Heat pump units are very similar to RTUs as both systems utilize a refrigeration system. However, heat pump units can use the refrigeration system to produce heat as the first stage. At colder weather conditions, auxiliary heat is turned on to provide additional heating capacity

A VFD can be installed on heat pump units as well. The schematic diagram is very similar to the RTU as shown in Figure 12.

3.4.3. Computer room air-conditioning units

A CRAC unit is one type of air-handling unit used for computer rooms or data centers. CRAC units are often located inside the data center and provide cooling to the servers. A typical CRAC unit includes direct expansion coil(s), compressors, supply fan(s), heater(s), and humidifier(s).

The traditional CRAC runs fans and compressors at a constant speed, which consumes extra fan and compressor power at partial load conditions. Application of a VFD converts a CRAC unit to a variable capacity CRAC unit. The VFD can be installed on the supply fan only or both the supply fan and compressors. The supply fan and compressor speeds are modulated to maintain the space temperature set point.

A data center is often a cooling-dominated building type. With the reduction of fan and compressor speeds, a great amount of fan and compressor power is saved.

3.4.4. Chillers

There are several ways to regulate system cooling capacity. (1) On–off control: This is the simplest way to control the capacity, but can cause frequent short cycling of compressors, which is detrimental to compressor performance. (2) Unloading control: This is often used for reciprocating compressors with multiple cylinders. (3) Slide valve control: This is often used for a screw compressor, which can adjust the compressor capacity with a wide range. (4) Hot gas bypass: This is not an efficient way as a mixture of hot and cold refrigerant is used. (5) Digital compressor: This is developed by Emerson for scroll compressors. The cooling capacity can be varied from 10% to 100%. (6) Variable speed compressor: This provides a smooth system capacity modulation with a wide range and is more energy efficient. Some chiller manufacturers already produced chillers with variable speed scroll compressors or variable speed screw compressors.

For a VFD-equipped compressor, the compressor speed is often modulated to maintain the supply water or supply air temperature set point.

4. VFD application on fault detection and diagnosis

Many researchers have studied fault detection and diagnosis (FDD) on HVAC systems. FDD technique is an effective way to improve the reliability of HVAC systems, and reduce the maintenance costs. There are a variety of methods and strategies on the equipment-level and system-level FDD, including AHUs, RTUs, etc. [19–21]. Almost all of the methods rely on the system operations measurements, such as temperature, humidity, pressure, airflow, and water flow.

Although VFDs are widely used in fans, pumps, and compressors in HVAC systems, most of these applications are focused on how to use the VFD to control motor speed. However, the VFD can measure several useful electrical-related parameters, which could be used for system monitoring and FDD purposes.

A typical VFD can measure and provide the output of speed/frequency, current, power, torque, and many other parameters. These electrical signals have inherent relationships with the system operating performances. For example, Li et al. [10] developed several fault signatures for a single-stage DX rooftop unit using fan power, compressor power, and supply air temperature measurements through an experimental study. With these known parameters, the components and system faults can be identified in advance. These signals can be sent out to an external controller or a BAS system through analog output signals or digital communication signals (Modbus, N2, FLN, BACNet, etc.).

Figures 13 and 14 present two configurations of connection between VFDs and the unit controller/BAS. In Figure 13, the VFD controls the speeds of multiple motors, such as fan motor, pump motors, or compressors. The controller monitors the operation of motors and receives motor operating information (such as speed, current, power, and torque) through digital communication. The controllers utilize this information and other system measurement readings (such as temperature) to perform FDD analysis.

It is also very common that each VFD controls only one motor, as shown in Figure 14. The controller communicates with each VFD and performs FDD analysis based on the operations of all motors.

One example is the application of the VFD on FDD in packaged RTUs. The common faults of an RTU include fouling evaporator coil, filter blockage, fouling condenser coil, refrigerant leakage, and improper charge. The common methods to detect these faults are using the measurement of multiple temperature and pressure points and comparing the actual readings to normal state readings. As a matter of fact, electrical signals can reflect the change of system performances. Recent research shows that the electrical signals, such as a VFD speed (frequency) and power, combined with other temperature parameters, can be used to detect these common faults based on experimental studies [10].

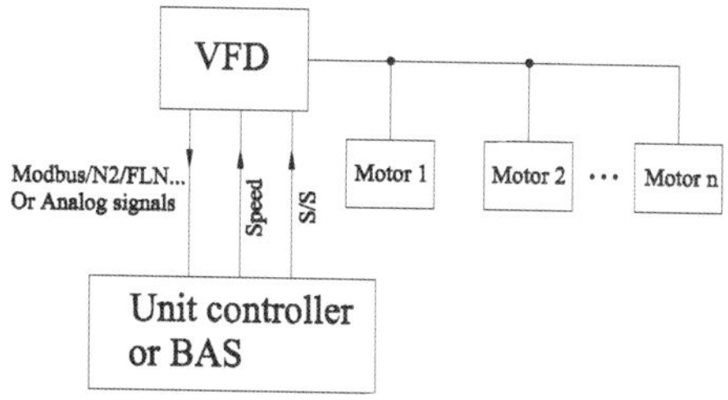

Figure 13. Configuration A showing the connection between the VFD and controller

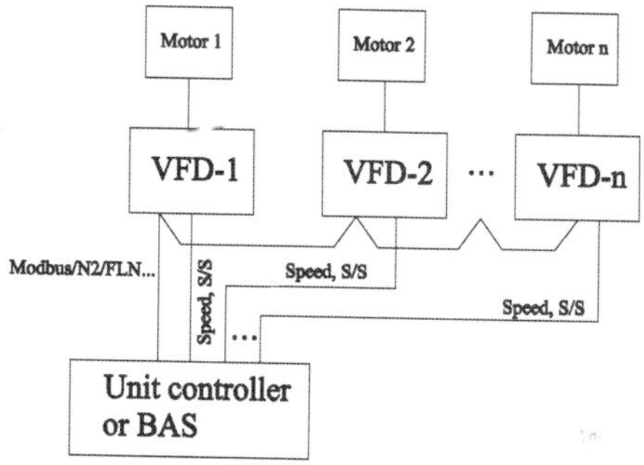

Figure 14. Configuration B showing the connection between the VFD and controller

To obtain the frequency (speed) and kilowatt for both the fan and compressor, both of them should be equipped with VFD, using a VFD to control both speeds or using dedicated VFD for fan and compressor. To monitor the performance of RTU, an outdoor air temperature (OAT) sensor and a supply air temperature (SAT) sensor are installed in the unit in addition to the VFD. The VFD speed and power are provided by the VFD itself and sent to an external controller or BAS through Modbus communication. The measured system parameters, such as VFD speed, VFD power, OAT, and SAT, are used to perform FDD on the existing RTU.

5. Application considerations

5.1. Minimum VFD speed

For all VFD applications, the maximum speed or frequency is relatively easy to set. In the United States, the maximum speed is usually 60 Hz. In some cases, a higher speed is used, which is not typical and recommended [22]. In contrast, the minimum speed setup needs more considerations because it has a potential impact on the building energy use and motor performance.

First, the motor itself has some limitations. VFD manufacturers often recommend a minimum speed of 30% of their rated speed (18 Hz) to prevent motor overheating due to inadequate airflow [23]. An inverter duty motor can have lower minimum setting as 20% (12 Hz). However, more considerations are needed to ensure effective operations.

For fans and pumps, the minimum speed can be as low as 6 Hz without creating motor overheat issue and other mechanical drawbacks [18]. Meanwhile, the operation factors should be considered as well, such as the indoor air quality (IAQ) requirements and air distribution requirements. If the fan speed is too low, with the same outdoor air damper position, less fresh air is delivered to the space. Therefore, a proper engineering calculation is needed. In addition, the operating mode places limitations on the minimum speed. For a single-zone unit running in cooling mode, a low speed could cause very low velocity at the outlet of ductwork, which may result in the cold air being dumped directly into space without a good mixture. In the heating mode, a speed that is too low may cause the hot air to stagnate on the upper level of space due to the buoyancy effect. Therefore, the actual minimum fan speed may be 20 Hz or so. In chilled water pump applications, the primary pump speed should be high enough to provide sufficient chilled water through chillers. Otherwise, the low-water-flow alarm could trip the operation of the chillers.

For compressors, their minimum speeds should be determined based on the oil return, as well as structural and safety requirements. For example, the manufacturer recommended a minimum VFD speed of 25 Hz for Discus compressors and 45 Hz for scroll compressors [24]. Most compressors have a vibration resonance issue at certain speeds. This can be solved by programming the VFD to skip this range, or by simply setting up a higher minimum speed to bypass this range.

5.2. Interferences

Most VFDs use pulse-width modulation to control the motor speed. PWM can create a large and rapid voltage swing, or an electromagnetic interference (EMI) because of the fast rise and fall times of the signals used by the PWM control circuits. The interference has adverse effects on the operation of the control system and motor components.

There are several recommendations to minimize interference from VFDs [25].

- Minimize the cable lengths between the VFD and motor. The longer the cable, the greater the potential for reflected voltage. The users shall follow the manufacturer's requirements for power cable installation. Generally, the cable length should be no more than 200 ft.
- Use the lowest VFD carrier frequency as it affects the maximum permitted cable length. The lower the frequency, the greater the maximum possible length of cable between the VFD and motors.
- Use an armored power cable. Metallic outer armor is recommended for the power cable to shield the system components from the high-frequency electric fields. Copper or aluminum should be used because steel does not provide effective shielding at high frequencies.

- Use separate metal conduits for input power, output power, control wires, and communication wires [26].
- Use isolation transformers for the VFD power. In this way, separate dedicated transformers and groundings are used for the VFD and control system. This grounding system could create a path to eliminate unwanted signals.
- Use other noise suppression components, such as input power filters, output power filters, and common mode chokes. These components can help suppress electrical noise in VFD applications.
- Select control-matched inverter-rated motors. These motors are designed to withstand the added stress when controlled by VFDs.

6. Summary

The VFD is an excellent electric device to control motor speed within the allowable operating range. The VFD applications on HVAC systems are presented in detail from the control perspective. Then, the application on FDD is introduced from the metering point of view. Lastly, the existing issues are summarized and recommendations are provided. Overall, VFDs play a great role in the optimal operation of building energy systems. The increased functionality and reliability along with the reduced cost make them more and more widely accepted and used in the HVAC industry. These applications will achieve a tremendous energy savings from motors.

Author details

Yunhua Li*

Address all correspondence to: yunhuali19@gmail.com

Bes-Tech Inc., Omaha, NE, USA

References

[1] Waide P, Brunner C.U.Energy-efficiency policy opportunities for electric motor-driven systems. OECD/IEA, International Energy Agency, Paris, France; 2011.

[2] Goetzler W, Sutherland T, Reis C. Energy savings potential and opportunities for high-efficiency electric motors in residential and commercial equipment. U.S. Department of Energy report [Internet]. 2013 Available from: http://energy.gov/sites/prod/

files/2014/02/f8/Motor%20Energy%20Savings%20Potential%20Report %202013-12-4.pdf [Access 2015-8-27]

[3] American Society of Heating, Refrigerating and Air-Conditioning Engineers. ASHRAE handbook: HVAC systems and equipment. Atlanta, GA: American Society of Heating, Refrigeration and Air-Conditioning Engineers.2008.

[4] Montgomery R, Mcdowall R.Fundamentals of HVAC control systems. Atlanta, GA: American Society of Heating, Refrigeration and Air-Conditioning Engineers; 2008.

[5] Itami T, Okoma K, Misawa K.. An experimental study of frequency-controlled compressor. In: International Compressor Engineering Conference at Purdue, Purdue University; 1982. p. 297–304.

[6] Qureshi T.Q., Tassou S.A. Variable-speed capacity control in refrigeration systems.. Applied Thermal Engineering. 1996;16(2):103–113.

[7] Cuevas C, Lebrun L. Testing and modelling of a variable speed scroll compressor. Applied Thermal Engineering. 2009;29(2–3):469–478.

[8] Ekren O, Sahin S, and Isler Y. Comparison of different controllers for variable speed compressor and electronic expansion valve. International Journal of Refrigeration. 2010;33 (6):1161–1168.

[9] Dardenne L, Fraccari E, Maggioni A, Molinaroli L, Proserpio L, Winandy E. Semi-empirical modelling of a variable speed scroll compressor with vapour injection.. Internatioal Journal of Refrigeration.. 2015;5(4):76–87.

[10] Li Y, Liu M, Lau J, Zhang B. Experimental study on electronic signatures of common faults for packaged DX rooftop units. Energy and Buildings. 2014;77:401–415.

[11] Li Y, Liu M, Lau J, Zhang B. Development of a variable speed compressor power model for single-stage packaged DX rooftop units. Applied Thermal Engineering. 2015;7(8):110–117.

[12] ANSI/ASHRAE/IESNA Standard 90.1-2010: energy standard for buildings except low-rise residential buildings. Atlanta, GA: American Society of Heating, Refrigeration and Air-Conditioning Engineers; 2010.

[13] Li Y, Liu M, Zhang B.VFD application in a manufacturing plant. Energy Engineering. 2013;110(2):58–77.

[14] Liu M, Claridge D.E., and Turner W.D. Continuous commissioning SM guidebook. Washington, DC: US Department of Energy; 2002.

[15] Liu M, Claridge D.E. The maximum potential energy savings from optimizing cold deck and hot deck reset schedules for dual duct VAV systems. Journal of Solar Energy Engineering. 1999;121(3):171–175.

[16] Hyman B.L., Bockmiller F. Primary chilled water loop retrofit. ASHRAE Journal. 2000;42(12):60–64.

[17] Liu M.. Variable water flow pumping for central chilled water systems. Journal of Solar Energy Engineering. 2002;124(3):300–304.

[18] Taylor S.T. Chilled water plant retrofit - a case study. ASHRAE Transactions. 2006;187–197.

[19] Seem J.E., House J.M. Integrated control and fault detection of air handling units. HVAC&R Research. 2009;15(1):25–55.

[20] Schein J, Bushby S.T. A hierarchical rule-based fault detection and diagnosis method for HVAC systems. HVAC&R Research. 2006;12(1):111–125.

[21] Li H, Braun J.E. A methodology for diagnosing multiple simultaneous faults in vapor compression air conditioners. HVAC&R Research. 2007;13(2):369–395.

[22] FLYGT. VFD recommendations [Internet]. 2015. Available from: http://www.flygtus.com/782832.pdf [Accessed: 2015-8-1]

[23] Tredinnick S. Variable speed pumping: how low can you go? [Internet]. 2006. Available from: http://www.districtenergy.org/pdfs/DEMagArticles/3Q06/3q06insideinsights.pdf [Accessed: 2015-8-1]

[24] Emerson Climate Technologies. Use of variable frequency drives (VFDs) with Copeland Scroll and Discuss compressors [Internet]. 2011. Available from: https://opi.emersonclimate.com/CPID/GRAPHICS/Types/AEB/ae1369.pdf [Accessed: 2015-8-1]

[25] Brown D, Harrold D, Hope R. Control system power and grounding better practice. London: Newnes; 2004.

[26] Frank J.B.Electromagnetic interference: what drive users need to know [Internet]. 2009. Available from: http://www.controleng.com/single-article/electromagnetic-interference-what-drive-users-need-to-know/f9c4ae94ed0d2381cf95ed2b914a6fd8.html [Accessed: 2015-7-10]